Quantum-Dot Cellular Automata Circuits for Nanocomputing Applications

This book provides a composite solution for optimal logic designs for Quantum-Dot Cellular Automata based circuits. It includes the basics of new logic functions and novel digital circuit designs, quantum computing with QCA, new trends in quantum and quantum-inspired algorithms and applications, and algorithms to support QCA designers.

Futuristic Developments in Quantum-Dot Cellular Automata Circuits for Nanocomputing includes QCA-based new nanoelectronics architectures that help in improving the logic computation and information flow at physical implementation level. The book discusses design methodologies to obtain an optimal layout for some of the basic logic circuits considering key metrics such as wire delays, cell counts, and circuit area that help in improving the logic computation and information flow at physical implementation level. Examines several challenges toward QCA technology like clocking mechanism, floorplan which would facilitate manufacturability, Electronic Design Automation (EDA) tools for design and fabrication like simulation, synthesis, testing, etc.

The book is intended for students and researchers in electronics and computer disciplines who are interested in this rapidly changing field under the umbrella of courses such as emerging nanotechnologies and its architecture, low-power digital design. The work will also help the manufacturing companies/industry professionals, in nanotechnology and semiconductor engineers in the development of low power quantum computers.

Materials, Devices, and Circuits: Design and Reliability
Series Editor: Shubham Tayal, K. K. Paliwal, Amit Kumar Jain

Tunneling Field Effect Transistors
Design, Modeling and Applications
Edited by T.S. Arun Samuel, Young Suh Song, Shubham Tayal, P. Vimala and
Shiromani Balmukund Rahi

Quantum-Dot Cellular Automata Circuits for Nanocomputing Applications

Edited by
Trailokya Nath Sasamal, Hari Mohan Gaur,
Ashutosh Kumar Singh, and Xiaoqing Wen

CRC Press
Taylor & Francis Group
Boca Raton London New York

CRC Press is an imprint of the
Taylor & Francis Group, an **informa** business

Designed cover image: QCA layout of 4-bit Parity Generator: Trailokya Nath Sasamal

First edition published 2024
by CRC Press
2385 NW Executive Center Drive, Suite 320, Boca Raton FL 33431

and by CRC Press
4 Park Square, Milton Park, Abingdon, Oxon, OX14 4RN

CRC Press is an imprint of Taylor & Francis Group, LLC

ISBN: 978-1-032-42018-9 (hbk)
ISBN: 978-1-032-42195-7 (pbk)
ISBN: 978-1-003-36163-3 (ebk)

DOI: 10.1201/9781003361633

Typeset in Times
by Newgen Publishing UK

Contents

Acknowledgement

We would like to thank Dr. Shubham Tayal, Dr. Krishan Kumar Paliwal, and Dr. Amit Kumar Jain Editors of the CRC Press/Taylor & Francis Books Series in Materials, Devices, and Circuits: Design and Reliability, for their kind invitation to perform this quite exciting and challenging project. We also thank Dr. Marc Gutierrez, from CRC Press/Taylor & Francis Group, for all the accorded technical support and kind facilities.

Our Sincere gratitude goes to all authors for their contributions without which this book could not have been edited. We emphasize here that all authors are deeply involved in research in the area of Quantum-Dot Cellular Automata, which provides an important impact on the contents of the corresponding chapters. We also thank all reviewers for their suggestions and illuminating views for each book chapter presented here in **Quantum-Dot Cellular Automata Circuits for Nanocomputing Applications.**

Editors Bio

Dr. Trailokya Nath Sasamal is currently working as Assistant Professor in the Department of Electronics & Communication Engineering at National Institute of Technology, Kurukshetra, India since August 2013. He has more than 11 years research and teaching experience in various University systems of India. Dr. Sasamal has obtained his Ph.D. degree from the Department of Electronics & Communication Engineering, NIT Kurukshetra, Haryana. He obtained his M. Tech degree in Electronics Engineering from Indian Institute of Technology, Banaras Hindu University, Varanasi, India. He obtained his B. Tech degree in Electronics & Telecommunication from the KEC, Bhubaneswar, India, in 2007. He has presented and published over 60 research papers in reputed journals and various national and international conferences. His research interests include Quantum-Dot Cellular Automata, Reversible logic, and new architectures for emerging nanodevices. He is the author of the book "Quantum-Dot Cellular Automata Based Digital Logic Circuits: A Design Perspective", published in Springer. He is also involved in reviewing processes in different journals and conferences such as; IEEE, IET, JCSC, IETE, DSJ, etc.

Dr. Hari Mohan Gaur is currently working with the School of Computer Science Engineering and Technology at Bennett University, Greater Noida, India. He obtained his Ph.D. from the National Institute of Technology Kurukshetra (NIT-KKR) in Reversible and Quantum Computation. Hari Mohan Gaur has more than 15 years of experience in academic, research and administrative capacities. He is a distinguished researcher, well known in Academic Fraternity for his interdisciplinary research in the areas of Quantum Computation, Fault Tolerant Digital Design, IOT and Data Security in Cloud Environment. Dr. Gaur holds the credit of contribution in several quality research journals of international repute published by IEEE, ACM, IET, Elsevier, etc. He is having a wide exposure of handling research proposals, international conferences, Training and Faculty Development Programs. He is involved in a joint research group involving eminent professors from top universities of US, UK, Japan, Taiwan, and Malaysia. He has also been editor and reviewer of several international journals and is a member of IEEE since 2017.

Prof. Ashutosh Kumar Singh is an esteemed researcher and academician in the domain of Electrical and Computer engineering. Currently, he is working as a Professor; Department of Computer Applications; National Institute of Technology; Kurukshetra, India. He has more than 20 years research, teaching and administrative experience in various University systems of the India, UK, Australia, and Malaysia. Dr. Singh obtained his Ph.D. degree in Electronics Engineering from the Indian Institute of Technology-BHU, India; Post Doc from Department of Computer Science, University of Bristol, United Kingdom and Charted Engineer from the United Kingdom. He is the recipient of Japan Society for the Promotion of Science (JSPS) fellowship for visit in University of Tokyo and other universities of Japan.

His research area includes Verification, Synthesis, Design and Testing of Digital Circuits, Predictive Data Analytics, Data Security in Cloud, Web Technology. He has more than 350 publications till now which includes peer reviewed journals, books, conferences, book chapters and news magazines in these areas. He has co-authored eight books including "Web Spam Detection Application using Neural Network", "Digital Systems Fundamentals", and "Computer System Organization & Architecture". Prof. Singh has worked as principal investigator/investigator for six sponsored research projects and was a key member on a project from EPSRC (United Kingdom) entitled "Logic Verification and Synthesis in New Framework".

Dr. Singh has visited several countries including Australia, United Kingdom, South Korea, China, Thailand, Indonesia, Japan, and USA for collaborative research work, invited talks and to present his research work. He had been entitled for 15 awards such as Merit Awards-2003 (Institute of Engineers), Best Poster Presenter-99 in 86th Indian Science Congress held in Chennai, INDIA, Best Paper Presenter of NSC'99 INDIA, and Bintulu Development Authority Best Postgraduate Research Paper Award for 2010, 2011, 2012.

Prof. Xiaoqing Wen (Fellow, IEEE) received the B. E. degree from Tsinghua University, China, in 1986, the M. E. degree from Hiroshima University, Japan, in 1990, and the Ph.D. degree from Osaka University, Japan, in 1993. From 1993 to 1997, he was an Assistant Professor at Akita University, Japan. He was a Visiting Researcher at the University of Wisconsin, Madison, USA, from October 1995 to March 1996. He joined SynTest Technologies, Inc., USA, in 1998, and served as its Chief Technology Officer until 2003. In 2004, he joined the Kyushu Institute of Technology, Japan, where he is currently a Professor and the Chair of the Department of Creative Informatics. He founded the Dependable Integrated Systems Research Center in 2015 and served as its Director until 2017. He has co-authored and co-edited two books: VLSI Test Principles and Architectures: Design for Testability (Morgan Kaufmann, 2006) and Power-Aware Testing and Test Strategies for Low Power Devices (Springer, 2009). He holds 43 U.S. patents and 14 Japanese patents on VLSI testing. His research interests include VLSI test, diagnosis, and testable design. He is a member of the IEICE, the IPSJ, and the REAJ. He received the 2008 IEICE-ISS Best Paper Award for his pioneering work on X-filling-based low-capture-power test generation. He has/is served/serving as an Associate Editor for the IEEE Transactions on Computer–Aided Design, the IEEE Transactions on Very Large Scale Integration (VLSI) Systems, and the Journal of Electronic Testing: Theory and Applications.

Contributors

Davood Aliakbari
ACECR institute of higher education, Isfahan branch, Isfahan, Iran

Aoran Cao
School of Computer Science and Technology, Anhui University, Hefei 230601, China

Deepak Garg
Department of Electronics and Communication Engineering, SRMIST, Delhi-NCR Campus, Ghaziabad, India

Hari Mohan Gaur
School of Computer Science Engineering and Technology, Bennett University, Greater Noida, India

Lakshminarayanan Gopalakrishnan
National Institute of Technology Tiruchirappalli, Tamil Nadu, India

Saptarshi Gupta
Department of Electronics and Communication Engineering, SRMIST, Delhi-NCR Campus, Ghaziabad, India

Vaibhav Jain
Department of Electronics and Communication Engineering, ABES Institute of Technology, Ghaziabad, India

Vineet Jaiswal
School of VLSI Design and Embedded System, NIT Kurukshetra, Haryana, India

Asghar Karimi
ACECR institute of higher education, Isfahan branch, Isfahan, Iran

Seok-Bum Ko
University of Saskatchewan, Saskatoon, Canada

Xuehua Li
School of Computer Science and Technology, Anhui University, Hefei 230601, China

Runqi Liu
School of Computer Science and Technology, Anhui University, Hefei 230601, China

Hamid Mahmoodian
ACECR Institute of Higher Education, Isfahan Branch, Isfahan, Iran

Sasan Ansarian Najafabadi
ACECR institute of higher education, Isfahan branch, Isfahan, Iran

Paramjeet
Department of Electronics and Communication Engineering, SRMIST, Delhi-NCR Campus, Ghaziabad, India

Marshal R
Indian Computer Emergency Response Team, Ministry of Electronics and Information Technology, New Delhi 110003, India

Abdalhossein Rezai
Department of Electrical Engineering, University of Science and Culture, Tehran, Iran

Saeed Ghorbani Rizi
ACECR Institute of Higher Education,
  Isfahan Branch, Isfahan, Iran

Gaurav Saini
Department of Electronics and
  Communication Engineering, NIT
  Kurukshetra, Haryana, India

Raja Sekar K
Centre for Development of Advanced
  Computing, Bengaluru, India

Devendra Kumar Sharma
Department of Electronics and
  Communication Engineering,

SRMIST, Delhi-NCR Campus,
  Ghaziabad, India

Ashutosh Kumar Singh
Department of Computer Applications,
  National Institute of Technology,
  Kurukshetra, India

Anantharaj Thalaimalai Vanaraj
Senior Technologist, R & D, Western
  Digital, California, USA

Aibin Yan
School of Computer Science and
  Technology, Anhui University, Hefei
  230601, China

1 Towards the Evaluation from Low Power VLSI to Quantum Circuits

Deepak Garg and Devendra Kumar Sharma

1.1 INTRODUCTION

The growing demand of portable systems and the requirement for limited power consumption in high-density ULSI circuits has led to rapid and constructive development in low-power design in recent years. The driving force behind these advancements is portable applications that require low power dissipation and high throughput, such as personal digital assistants, notebook computers, portable communication devices. Low-power output of integrated digital circuits has emerged as the most efficient and fast-growing field of CMOS design. Mobile phones, microprocessors, laptops, and other high-performance digital applications are among the many battery-operated portable devices and applications where the requirement for low-power design is becoming a significant problem. High operating speeds and increasing chip density result in the design of extremely complex chips with high clock frequencies. When a chip's clock frequency rises, the chip's power consumption and consequently its temperature rise linearly. Since the finished heat has to be effectively removed to keep the chip temperature in a tolerable condition, the cost of packaging and heat removal becomes an important and useful factor. Improvements in scaling with decreased threshold and supply power voltages lead to improvements in the leakages in MOS transistors. Numerous studies have demonstrated that leakage power is 40% of the total power consumed in nanotechnology [1].

The authenticity of the ULSI is yet another concern that illustrates the requirement for low-power circuit structure. Digital circuit's peak power dissipation and reliability problems like electro migration and power carrier destruction are intimately related to one another. Also, heat stress caused by the heat transfer to the chip is a major dependency problem. As a result, the reduction in power consumption is also useful in order to increase the reliability. Various researchers give different ideas from device level to building level to overcome the power consumption problem. Delay, area, and power trade-offs cannot be completely avoided, though. Therefore, designers are needed to choose a suitable technique that satisfies product use and requirements [2]. Techniques used to achieve high speed and low power consumption for use in digital systems take a large area, from device/ process level to algorithm level. The important

DOI: 10.1201/9781003361633-1

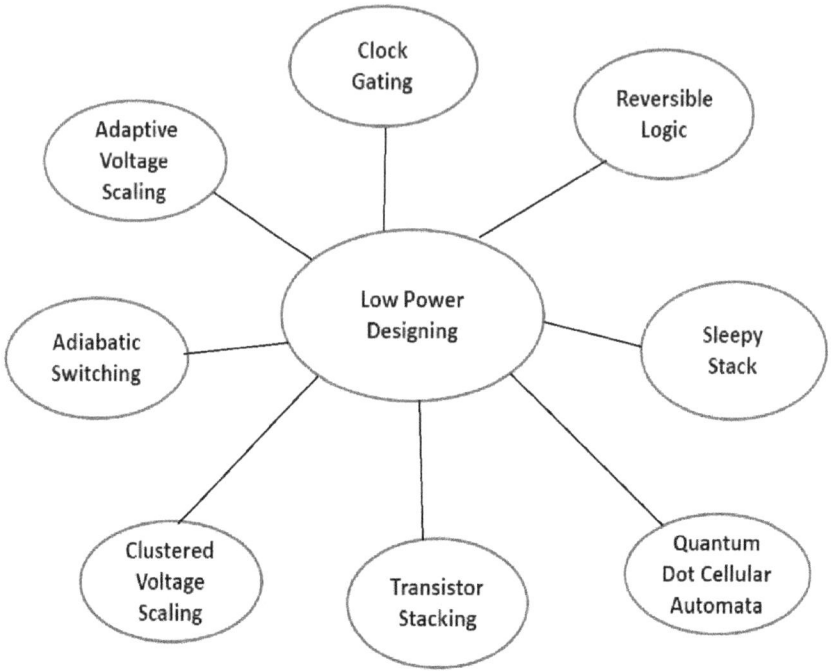

FIGURE 1.1   CMOS inverter for dynamic power reduction.

factors for reducing power consumption for any device are threshold voltage, device geometries and connection properties. To decrease the power loss at the transistor level and circuit-level methods can be utilized, such as the right selection of circuit design patterns, and clocking techniques. The system's power consumption can be decreased by carefully choosing data processing methods, notably by reducing the count of switching events required to complete a given task.

Figure 1.1 shows the different methods for dynamic power reduction for CMOS inverter.

Device scaling is severely constrained by power dissipation as a result of the development in VLSI design technology, which offers exponentially decreasing device dimensions and exponentially increasing circuit complexity. This calls for better power optimization techniques, such as reversible logic. Quantum wires connect the quantum gates that make up a quantum circuit. The unitary transformation determines the actual structure of a quantum circuit, the number and types of gates, as well as the connecting scheme. Modern chip architecture software frequently focuses on power optimization today. By dramatically lowering the power needs, reversible logic is changing the entire situation. By preventing power loss owing to information loss as in irreversible processes, it ensures almost energy-free computation [3]. To push the limits of logic design with extremely low power consumption, researchers are making impressive efforts to build combinational and sequential logic functions.

In this chapter, we discuss circuit and logic design techniques to reduce short circuit, leakage, and dynamic power dissipation.

1.2 NEED OF LOW POWER DESIGN

Recent years have seen significant and creative advancements in low-power design due to the growing importance of portable systems and the requirement to restrict power consumption in very-high density VLSI processors. Portable applications including notebook computers and personal digital assistants that need low power dissipation and high throughput are what are driving these innovations. In the majority of these situations, it is necessary to achieve both the challenging objectives of high throughput and high chip density in addition to the low power consumption requirements. As a result, the field of CMOS design for low-power digital integrated circuits has become quite active and is developing very quickly.

The whole power consumption of the portable system is often subject to very rigorous requirements due to the restricted battery lifetime. Even while new types of rechargeable batteries, such nickel-metal hydride (NiMH) batteries, are being created with energy capacities that surpass those of the more common nickel-cadmium (NiCd) batteries, a revolutionary rise in energy capacity is not anticipated anytime soon. The current battery technologies (such NiMH) offer an energy density of roughly 30 Watt-hour/pound, which is still underwhelming given the growing number of applications for portable devices. Therefore, a significant problem in the design of portable systems is lowering the power dissipation of integrated circuits through design enhancements.

In high-performance digital circuits, such as microcontrollers, DSPs, and other advanced applications, the requirement for low-power design is also turning into a significant problem. High clock frequencies are a result of the construction of extremely complicated processors due to increasing chip density and operating speed. The chip's power dissipation and consequent temperature rise linearly as the chip's clock frequency rises. Since the dissipated heat must be removed effectively to keep the chip temperature at a reasonable level, the cost of packing, cooling, and heat removal becomes a major consideration. The Intel Pentium, DEC Alpha, and PowerPC, among other high-performance microprocessor chips created in the early 1990s, operate at clock rates between 100 and 300 MHz and typically consume between 20 and 50 W of power. Another issue that highlights the necessity for low-power design is ULSI reliability. Peak power dissipation of digital systems and reliability issues like electro migration and hot-carrier driven device deterioration are closely related. A significant reliability issue is the thermal stress brought on by the heat dissipation on chips. So, cutting back on energy use is equally essential for improving reliability.

Low power consumption in digital systems can be achieved using a variety of approaches, ranging from device/process level to algorithm level. Reduced power consumption is mostly a result of device parameters (such as threshold voltage), device geometries, and connection qualities. To decrease power dissipation at the transistor level, circuit-level techniques can be utilized, such as the appropriate selection of circuit design strategies, lowering the voltage swing schemes. Smart power management of different system blocks, the use of parallelism and pipelining, and the design of bus topologies are examples of architecture-level controls. Finally, by carefully choosing the data processing algorithms, specifically to limit the amount of switching events for a specific activity, the system's power consumption can be decreased.

1.3   POWER CONSUMPTION IN CMOS

Low power design means the capacity to decrease all three components of power consumption to CMOS systems during VLSI product development. Dynamic power, short circuit power, and leakage power are added to determine a CMOS circuit's overall power usage.

$$P_{Total} = P_{Dynamic} + P_{Short} + P_{Leakage}$$

1.3.1   DYNAMIC POWER

The circuit-driven charging and discharging of load capacitors generates dynamic power, also known as active power [4]. The most accepted method to optimization of the power is the supply voltage scaling, since it normally yields considerable power saving due to the quadratic dependence of dynamic power on power supply VDD. The main disadvantage of this approach is that lowering the power supply voltage affects the speed of the circuit. Therefore, the design and technology solution is required to enhance the circuit performance which can be reduced by the lower voltage. Figure 1.2 depicts the dynamic power dissipation for a CMOS inverter circuit.

The dynamic power is given by equation (1).

$$P_{Dynamic} = \alpha * C_{Load} * V_{DD}^2 * f * N_{SW \, (1)}$$

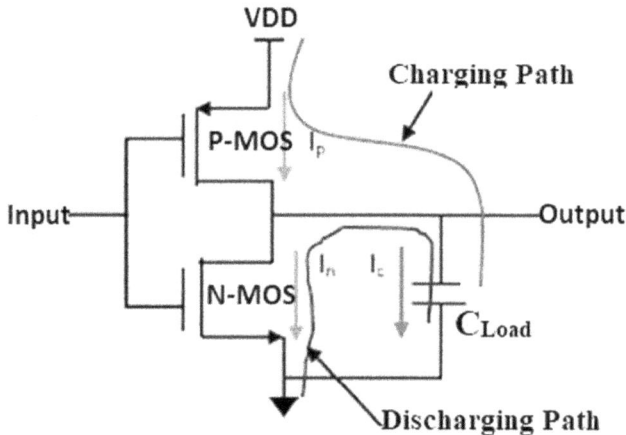

FIGURE 1.2   CMOS inverter for dynamic power reduction.

where,
α = Switching Activity
C_{Load} = Load Capacitance
V_{DD} = Supply Voltage
f = Clock Frequency
N_{SW} = Number of switching bits

1.3.2 LEAKAGE / STATIC POWER

The portion of power usage that is unaffected by activity is known as static power. The power used by the transistor when it is off owing to reverse bias current is known as leakage power. Power used by a device that is unrelated to state changes is known as leakage power [4]. The foremost challenge with this energy is for the duration of inactive state of the tool. Leakage power is the power which needs to be focused extra in deep submicron layout of the tool because it exponentially relies upon on length of the tool. Various reasons of leakage power are represented in Figure 1.3, consisting of the opposite bias modern-day, the channel leakage modern-day beneath the threshold, the perforation, the drain leakage caused via way of means of the grid, the slender width effect, the tunnel oxide grid glide, and the new assist injection glide [4].

Where,
I_1: punchthrough
I_2: weak inversion (subthreshold channel leakage)
I_3: narrow-width effect
I_4: Gate-Induced Drain leakage (GIDL)
I_5: PN reverse-bias current
I_6: Drain Induced Barrier-Lowering (DIBL) effect
I_7: hot-carrier injection
I_8: gate-oxide tunneling

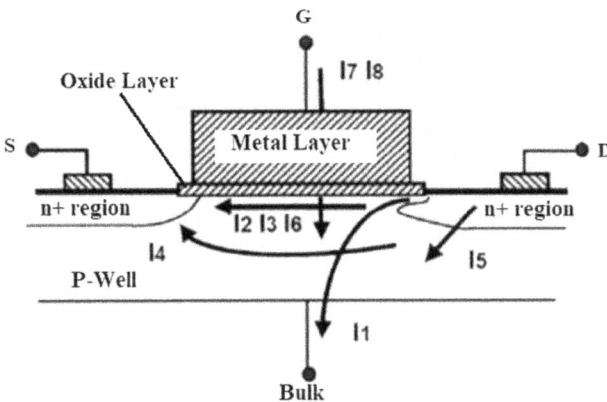

FIGURE 1.3 Leakage power reasons.

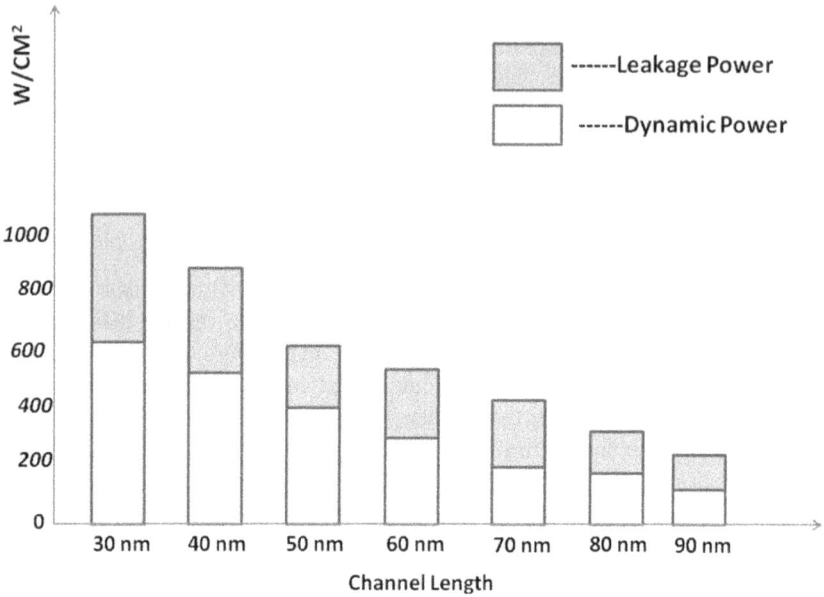

FIGURE 1.4 Leakage power and dynamic power at various deep submicron CMOS technologies.

G: Gate terminal
S: Source terminal
D: Drain terminal
In addition, even in standby, as shown in Figure 1.4, significant leakage power will dramatically increase dependability-related issues.

1.3.3 SHORT CIRCUIT POWER

Short circuit power is the power exhaust during the change in the output level of a CMOS logic gate. In the circuit, when both transistors *viz* NMOS and PMOS are activated then short circuit current is generated during signal transitions and there is a direct path between ground and power supply (V_{DD}). The short circuit power reduction for a CMOS inverter circuit is shown in Figure 1.5 [15].

Principles of low power designs are given as:

1) High frequency device is used at the lowest possible quantity.
2) Disconnects the power supply when the system is not working.
3) Controlling power dissipation at architectural level.
4) Using pipelining and parallelism to achieve the lowest necessary operating frequency using the lowest supply voltage feasible.

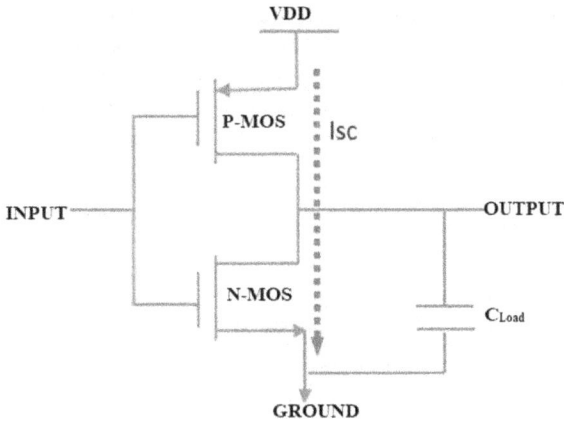

FIGURE 1.5 CMOS inverter for short circuit power reduction.

1.4 DIFFERENT LEVELS OF ABSTRACTION OF CUSTOMIZATION

The design levels of the integrated low-power techniques need to be optimized. The different levels of abstraction are discussed in the subsequent sub sections.

1.4.1 SYSTEM LEVEL

The topic of system level power optimization was discussed by Tong *et al.* [5]. The process of transferring a high-level system model to the architecture is known as system-level design. To obtain about 75% performance optimization, different architectures and processing algorithms with suitable executable specifications are required. Adaptive voltage scaling (AVS), branch reduction, hardware merging, loop unwinding and merging, preferred loop unwinding, and memory access reduction are a few efficient methods used to fine-tune system performance.

1.4.2 ALGORITHM LEVEL

According to Sharma *et al.* [6], electricity intake at algorithm degree relates the right desire of set of rules, phrase length, modular interfaces, generation implementation, hardware and software program selection, and behavioral constraints and trade-off will reduce the required electricity. It is a totally beneficial set of rules with a small range of operations as it would require a small hardware. By growing concurrency, we are able to grow the effectiveness of that device.

1.4.3 ARCHITECTURE LEVEL

According to Sharma *et al.* [6], the effects of low power approaches may be more significant at the architecture level than at the gate level. He emphasized the power management, pipeline, and distributed methods that might be used to decrease the facility dissipation at the design level. Similarity and pipelining are two strategies that can optimize power and space while maintaining an equivalent throughput. The combination of pipelining and parallelism may provide the more power reduction, as he indicated via various frequency and voltage islands, decrease in change activity, and the implementation of thorough logic transformations techniques at the design level to decrease the power consumption.

1.4.4 GATE LEVEL

At the gate level, we gain an accurate verification of power usage, according to Babu *et al.* [7]. By using clock gating, power gating, and clock tree optimization approaches, up to 20% of power can be saved. We can use logic level changes at the gate level as well to decrease switching activity and hence power consumption.

1.4.5 TRANSISTOR LEVEL

Power dissipation and gate delay in a combinational logic circuit are influenced by transistor size. A logic gate with wider transistors will have a decreased gate delay, but its switching power dissipation will be higher. A transistor with a shorter channel length will have higher subthreshold leakage, which increases the static power dissipation. On the other hand, due to a larger saturation current, short channel transistors are faster. According to Gupta *et al.* [8], sophisticated processes can be used to create transistors with various threshold voltages. Using a variety of CMOS transistors with different threshold voltages can reduce power consumption by up to 30%. There are two distinct thresholds, referred to as high threshold V_{th} and low threshold V_{th}. High threshold transistors can be utilized in non-critical circuits despite being slower and having less leakage. Figure 1.6 illustrates the specifics of the power savings, speed, and error trade-off at various levels of abstraction.

1.5 POWER OPTIMIZATION METHODS

Power optimization methods are very important and useful for the growth of portable devices. It is the process of using electronic design automation tools to decrease a digital design's power usage while maintaining functionality.

1.5.1 DYNAMIC POWER REDUCTION METHODS

For dynamic power reduction in VLSI circuits, various designers use different power optimizing techniques at different levels of designing. At this time we will discuss some useful power reduction methods, i.e. Clock gating, Dynamic Voltage and Frequency Scaling (DVFS), Clustered Voltage Scaling (CVS), Dual V_{DD}, Adaptive Voltage Scaling (AVS).

1.5.1.1 Clock Gating

One of the most effective and widely used methods for reducing the power of the clock signal is clock gating. We can mask the clock signal to these gating modules

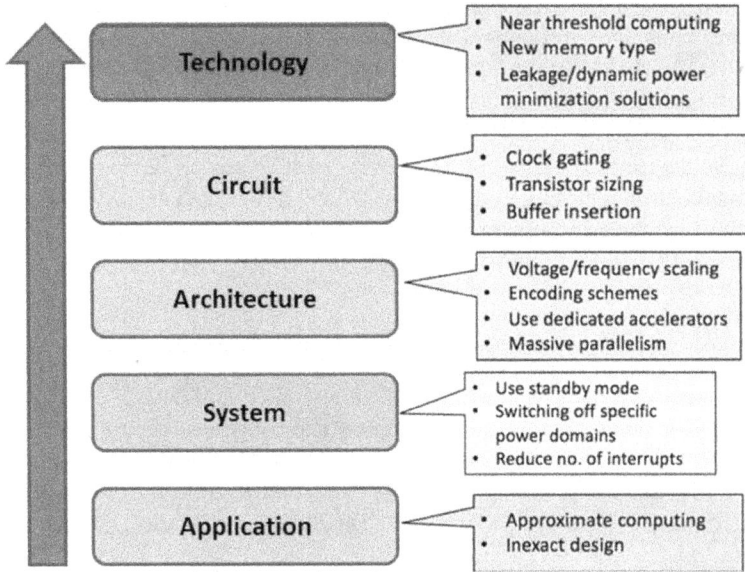

FIGURE 1.6 Energy or power optimization methods at various levels of abstraction.

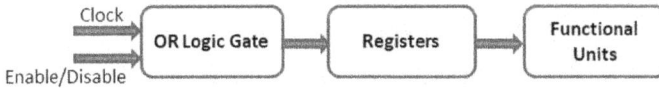

FIGURE 1.7 Clock gating method.

when a clock signal to the functional block is not required for a long time. NOR and NAND gates are typically used to block the clock signal from reaching functional units. The amount of power consumed by a circuit normally depends on the signal transition switching. Mainly clock pulse is responsible for any signal activities. Clock is the brain of the whole system so wherever clock transition takes place all the circuit operates synchronously. Sometimes we don't need that clock to some blocks so if we disable this clock to that particular block then switching activity is decreased and activity factor α will also decrease and power dissipation will also be decreased.

The clock loads the inactive registers indirectly, via the OR gate with the help of the enable signal. When we are certain that a functional unit is not needed, we set the enable to 1 to ensure that the output of the OR gate is always 1, the value of the register remains constant, and there is no signal transition into a functional unit. Depending on whether the functional unit is required or not, additional logic is needed when switching off the clock signal from the functional unit. However, additional circuitry (OR gate and for enabling various circuitry) may cause the clock signal to add some delay to the critical path; in this case, skew analysis is needed. Figure 1.7 illustrates the clock gating technique using OR gate.

Because this technique is intended for positive edge, the clock and dynamic power are maintained throughout the clock's negative edge. The controlling device clock is OFF if the target device clock is ON, and vice versa. By using synchronous circuits, this technology decreases hardware complexity and can save up to 20% on power. A good or terrible lock can also be utilized to conserve energy. By conserving energy, the device clock remains "ON" even while the target device clock is "ON". Additionally, the device clock command is "OFF" if the target device clock is "OFF". By minimizing the unnecessary changes of the clock network, more power may be saved in this [9].

1.5.1.2 Dual V_{DD}

In the Dual V_{DD} architecture [14], the supply voltages for the routing blocks and logic are programmed to use less power by assigning low-V_{DD} to design channels that are not timing-critical and high-V_{DD} to timing-critical pathways. On the other hand, the static current flows at the interface between the low-V_{DD} and high-V_{DD} parts whenever two different source voltages coexist. Using a level converter, a low-V_{DD} can be boosted to a high-V_{DD}. Figure 1.8 depicts a Dual V_{DD} Configuration Logic Block [24].

1.5.1.3 Clustered Voltage Scaling (CVS)

With the help of clustered voltage scaling (CVS), power consumption can be decreased without having to change circuit operation across two supply voltages [10]. As depicted in Figure 1.9, the critical path (represented by arrow direction) gates operate on the lower supply to decrease power consumption. To decrease the count of interface level shifters required, circuits that operate at reduced voltages are bundled, resulting in bundled voltage scaling. Here, a voltage transition along a path is only permitted once, and level conversion only occurs at flip-flops. In Figure 1.9, the input signals and output signals of a logic circuit are represented by $I_1 \ldots I_5$ and $O_1 \ldots O_5$ respectively.

1.5.1.4 Multi-voltage (MV)

MV concerns the operation of various areas of a design at various voltage levels [11]. The increased voltage offered is only connected to the specific areas that need it to satisfy performance goals. Other components of the design operate at a lower voltage, which saves a lot of energy. Multiple voltage is often a method to decrease the dynamic power, but lower voltage levels also reduce leakage power.

1.5.1.5 Dynamic Voltage and Frequency Scaling (DVFS)

A possibility for low-power VLSI design, dynamic voltage, and frequency scaling uses the quadratic relationship between supply voltage and circuit power usage to increase the overall energy utilization. This scaling method scales the targeted power domain's voltage levels into fixed discrete voltage steps. Many circuits have time-varying performance requirements, according to R. Sivakumar et al. [12], the facility can be saved by decreasing the clock frequency to a level that will allow the task to be finished on time before decreasing the voltage to that level. This can be referred to as dynamic voltage frequency scaling (DVFS). This system can be enforced by

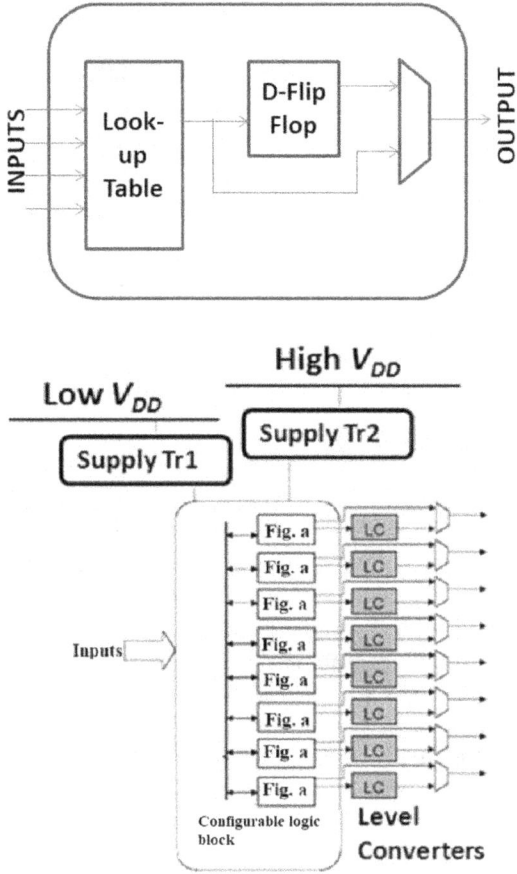

FIGURE 1.8 Dual V_{DD} configuration logic block.

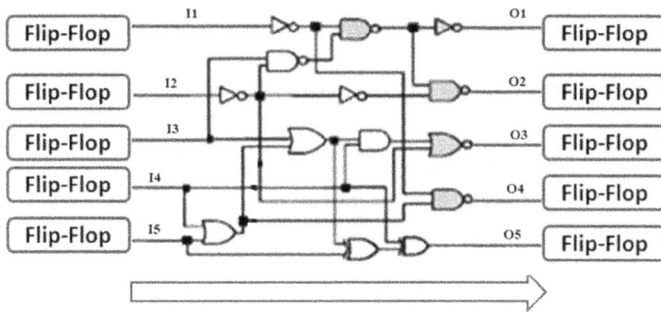

FIGURE 1.9 Critical path gates operate at lower supply levels.

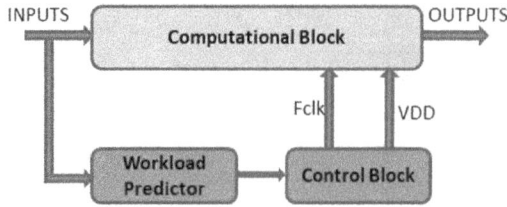

FIGURE 1.10 Frequency scaling and dynamic voltage approach block diagram.

correct dominant program [12]. The modification of the operational voltage and / or frequency that a tool operates at, throughout its operation, specifies that the minimum voltage and / or frequency needed for the right operation of a particular mode is called DVFS, dynamic voltage and frequency scale. [13].

DVFS allows devices to perform necessary activities with the least amount of required power. To maximize power savings, battery life, and device lifetime while yet retaining immediate computing performance available, the technology is used in basically all modern computer hardware. Figure 1.10 displays the block diagram for the frequency scaling and dynamic voltage approach.

1.5.1.6 Adaptive Voltage Scaling (AVS)

Adaptive voltage scaling (AVS) is a closed-loop control system that can decrease the overall consumption of power by up to 60%. Using a control system technique, adaptive voltage scaling delivers the lowest operating voltage for a specific process frequency [16]. By automatically altering the output voltage of the ability to provide compensation for process and temperature variation within the CPU, the AVS loop controls the processor's performance [17]. The AVS loop also lessens the power supply's tolerance. AVS utilizes up to 45% less energy than techniques for open loop voltage scaling, such as Dynamic Voltage Scaling. AVS is a system-level schematic that includes both a CPU and components for electrical power. The Advanced Power Controller, which remains on the processor, manages the AVS loop. The power supply's slave power controller (SPC), which translates the APC instructions, is located there. The APC and SPC were given IP, which handled the handshaking for the frequency and voltage ratings automatically, making it easier to integrate the system into the application.

1.5.2 Static Power Reduction Methods

There are several strategies for reducing power dissipation depending on the phases of operation (Active, Standby). We can reduce this power by using techniques like technology scaling, clock frequency scaling, voltage scaling, switching activity reduction, etc.

1.5.2.1 Multi Threshold CMOS Optimization (MTCMOS)

According to [18], the MTCMOS technology is highly useful for minimizing leakage currents in the CMOS standby mode. To maximize power and delay, this method makes use of transistors with different threshold voltages. To reduce clock durations, low voltage devices were used in this instance in important delay lines. Higher voltage devices were utilized on noncritical channels without incurring a delay penalty to lower the static leakage power. Both low V_{TH} and high V_{TH} MOS transistors are used in the CMOS inverter circuit with MTCMOS technology. In the active mode of the CMOS inverter circuits both sleep transistors (NMOS & PMOS) are in ON state connecting V_{DD} and ground to the circuit. To benefit from greater speed, both of these transistors have low V_{TH} values. In the standby mode of the CMOS inverter circuit both sleep transistors are in OFF state to isolate V_{DD} and ground from the circuit. To reduce leakage current, both of these transistors have high V_{TH} values.

1.5.2.2 Transistor Stacking

Leakage power in active mode can be reduced using the transistor stacking technique. By stacking two OFF transistors, as opposed to using only one, sub-threshold leakage is greatly reduced [19]. The influence of sub-threshold leakage current (Isub) on source voltage is used in transistor stacking (Vs). Sub-threshold leakage current exponentially decreases with increasing transistor source voltage. For this process, three transistors T1, T2 and T3 with the channel width of W, W/2 and W/2 respectively are used. When the circuit does not naturally stack transistors, we employ transistors with a width of W/2 in place of W to take advantage of the stacking effect. As seen in Figure 1.11, this process is known as forced stacking.

1.5.2.3 Sleepy Stack Approach

The term "sleepy stack method" refers to the fusion of the stack technique in standby mode with the sleep transistor approach in active mode [20]. Forced stacking is used first in this strategy. The next step is to add a sleep transistor to one of the

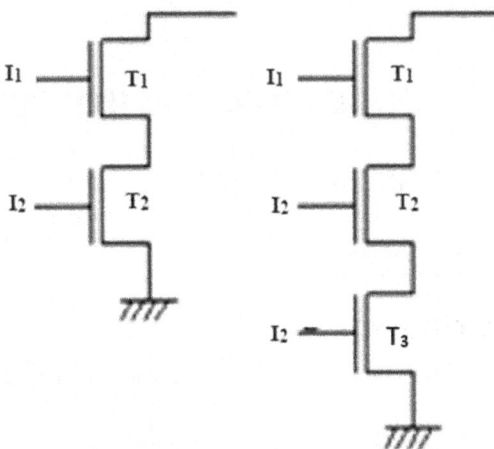

FIGURE 1.11   Circuit of forced stacking.

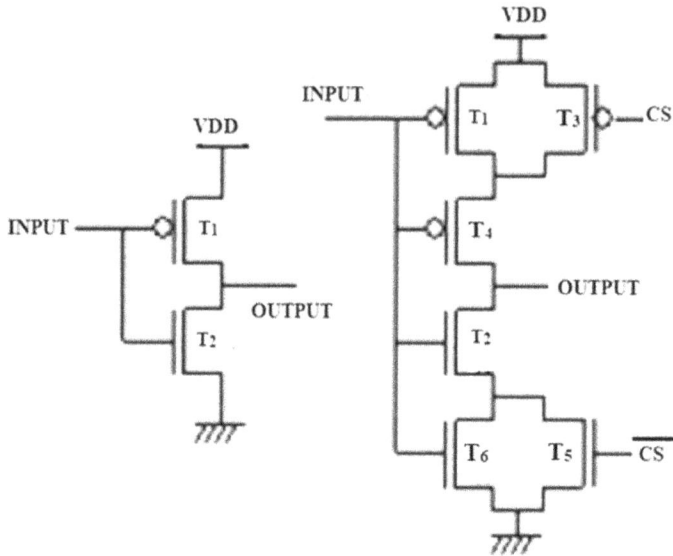

FIGURE 1.12   Circuit of sleepy stack inverter.

parallel-stacked transistors. As a result, the stacked transistor stops the power leak while the inactive transistor switches off in standby mode. Compared to the force stacking technique, this results in a propagation delay during active mode that is minimized. The sleep transistor is disabled while the stacked transistor stops power leaks during standby mode. Figure 1.12 depicts a sleepy stack inverter's circuit diagram. Three PMOS transistors T1, T3 and T4 and Three NMOS transistors T2, T5 and T6 having channel width of W and W/2 respectively are utilized in this technique. In Figure 1.12, CS and CS' signals represent the sleep control signals respectively.

1.6   ADVANCED POWER OPTIMIZATION TECHNIQUES

Nowadays, numerous cutting-edge power optimization strategies are applied in low power circuit design by various researchers. Here, we'll talk about three of the most effective power-saving methods: adiabatic switching, reversible logic, and Quantum-Dot Cellular Automata.

1.6.1   ADIABATIC SWITCHING

Another approach to the power dissipation decrease in digital logic is adiabatic switching. Adiabatic allows for the reuse of energy from the load capacitor as opposed to sending the load to the earth and using that energy. Adiabatic logic, which uses dynamic logic to save energy, is also known as Charge Recovery Logic or Charge Recycling Logic.

The operation of adiabatic circuits is governed by a few fundamental criteria, such as:

1) When there is a voltage potential between the drain and source terminals, a transistor should never be turned on.
2) The voltage across no transistor should ever be abruptly changed [21].

Adiabatic logic operates on the principle of switching power-consumption-inducing activities by returning stored energy to the supply. Therefore, a low-power VLSI circuit that employs reversible logic is referred to as having adiabatic logic. The timing of the clock pulse is essential to the operation of an adiabatic logic circuit. The user can realize the two primary design principles for the adiabatic logic circuit throughout each phase of the clock pulse. Resetting the energy on the power clock during the recovery period will result in significant energy savings [22]. Contrary to conventional non-adiabatic systems, which employed constant voltage charging from a set power capacity, logical adiabatic circuits' power supply used constant current charging (or something close to it). Circuit components with the ability to store energy have also been used as power supplies for adiabatic logic circuits. Adiabatic logic circuit is classified into the following types:

I. Partially Adiabatic Logic
II. Fully Adiabatic Logic.

1.6.1.1 Partially Adiabatic Logic

Some charge gets transferred to the ground in this type of adiabatic logic. Due to this, a portion of energy can only be recovered. Compared to a completely adiabatic logic circuit, this adiabatic logic circuit is simpler to implement. Partially Adiabatic Logic circuit is classified into following types:

I.   Positive Feedback Adiabatic Logic (PFAL)
II. Clocked Adiabatic Logic (CAL)
III. Efficient Charge Recovery Logic (ECRL)
IV. 2N-2N2P Adiabatic Logic.

1.6.1.2 Fully Adiabatic Logic

All charge on the load capacitance gets recovered and feedback to the power supply. In comparison to a partial adiabatic logic circuit, this adiabatic logic circuit is slower and more complex in design. Fully Adiabatic Logic circuit is classified into following types:

I.   Spilt-Level Charge Recovery Logic (SCRL)
II. Phase Adiabatic Static Clocked Logic (PASCL)
III. Pre-resolve and Sense Adiabatic Logic (PSAL)
IV. Pass-transistor Adiabatic Logic (PAL).

1.6.2   QUANTUM-DOT CELLULAR AUTOMATA

A recently created paradigm for digital design called Quantum-Dot Cellular Automata (QCA) has the potential to replace the current CMOS technology. The QCA technique relies on electrical communication between the cells rather than current transfer. The QCA cell essentially consists of metal islands or quantum dots that are spaced apart by a particular amount. Information is sent entirely through interactions between these localized electrons in potential wells. QCA is an innovative computing paradigm that is well suited for scaling to molecular dimensions and implementation in nanoelectronic systems.

The Quantum-Dot Cellular Automata proposed by Lent [23] is one of the prospective alternatives to CMOS technology. It proposes the idea of employing nanoscale structures, such as quantum dots, to build cellular automata. The physics of the interactions between the electrons positioned in the potential wells is the basis of QCA. One option for nanostructures to be employed in QCA is quantum dots. The width of a quantum dot, a small semiconductor crystal, is around a thousand times smaller than that of a human hair. QCA technology retains a strong position as an alternative to CMOS technology because of the benefit of ultra-low size [25]. There are a limited number of these quantum dots or metal islands that make up the fundamental QCA cell.

The coulombic interaction and quantum mechanical tunneling between the electrons concentrated in the wells serve as the foundation for the QCA's operational mechanism. The Coulomb's rule, which states that due to coulombic repulsion, electrons tend to occupy places with maximum spacing, governs how the electrons align in a QCA cell [26]. By taking into account the potential alignments of electrons in the cell, this can be logically comprehended. Due to the square shape of the QCA cell, Figure 1.13 depicts the two potential polarizations.

The basic building blocks of QCA-based circuits are called QCA cells. A fundamental QCA cell has four quantum dots at each corner of a square structure that are thought of as four potential vacancies or areas for the electrons to go, as seen in Figure 1.13. Two free electrons are available in a QCA cell, which can shift between the dots within the cell. There are two possible polarizations based on the electron's position. There are two conceivable ground states with polarizations of "1" and "−1", which stand in for binary "0" and "1", respectively. These two QCA cell states are depicted in Figure 1.14 [28].

1.6.2.1   Present Scenario of QCA Technology

The Quantum-Dot Cellular Automata technology is a major player in the post-CMOS future. It has significant effects on circuit performance and design, particularly when

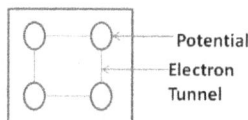

FIGURE 1.13   Schematic structure of a QCA cell.

Logic 0

Logic 1

Quantum Dot

Electron

Electron

Quantum Dot

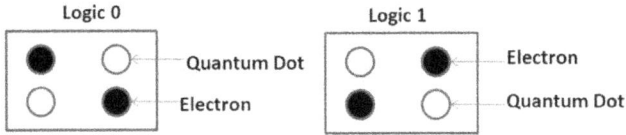

FIGURE 1.14   QCA cell as binary '0' and '1'.

feedback is present in sequential circuits. The circuit, which uses realistic protein alignment sequences as a test bench, is described and simulated with HDL language and is based on nanomagnetic logic as a QCA implementation. It is designed down to the layout level while taking technological limitations and experimentally verified structures into consideration. The results present represent a significant development in the area of emerging technologies because:

I.   They are established on a quantitative methodology and rely on a realistic, complicated circuit made up of a wide range of QCA blocks.
II. They are strictly calculated based on the technology limitations of the time without making irrational assumptions.
III. They offer a set of general guidelines for designing difficult sequential circuits using intrinsically pipelined technology, such as QCA.
IV. They demonstrate how to maximize the performance of the circuits using a real application benchmark.

1.6.2.2   Advantages of QCA over CMOS Technology

i.   QCA technology has ultra-low circuit size as compared to CMOS technology for performing the same operation.
ii. QCA technology has higher performing speed as compared to CMOS technology.
iii. QCA technology has highly scalable circuits as compared to CMOS technology.
iv. QCA technology has higher switching frequency as compared to CMOS technology.
v.   QCA technology has low power consumption as compared to CMOS technology circuits.

1.6.2.3   Applications of QCA

The well-known CMOS technology might soon be replaced by QCA technology. It provides significant advantages over other nanotechnology at the base level. The complete world will transform if QCA technology can be properly fabricated. All devices will shrink in size and become capable of achieving extraordinarily high speeds while using very little power to operate. Additionally, they will be charged independently utilizing tiny solar cells constructed of QCA cells. QCA microprocessors will be there in every field of technology and they will work at a bottom level without their presence to be noticed. Every aspect of technology will be impacted by QCA technology, including:

i.   Security
ii. Military
iii. Software
iv. Communication
v.   Satellites and Radars
vi. Navigation
vii. Gaming Industry
viii. Small scale of FPGA, CPLD and ASIC chip design industry
ix. Big data and Artificial Intelligence industry.

The world will actually be ruled by QCA technology in the near future.

1.6.3 REVERSIBLE LOGIC

Reversible logic has drawn a lot of interest recently because of its capacity to lower power dissipation, which is a crucial requirement in low power VLSI design. It has numerous uses in nanotechnology, DNA computing, low power CMOS and optical information processing, and quantum computing. Since the information contained in the circuits can be retrieved in the event of a loss, reversible logic is also known as information lossless logic.

1.6.3.1 Reversible Logic Gate

Reversible gate, also known as reversible logic gates, has an equal number of inputs and outputs. The amount of energy lost during computations will be minimized when the number of inputs and outputs is equal. An n-input, n-output logic device with one-to-one mapping is a reversible logic gate. This helps in separating the outputs from the inputs and enables the recovery of the inputs' individual identities from the outcomes. Direct fan-out is also not permitted in the synthesis of reversible circuits since the one-to-many idea is not reversible. Fan-out is however accomplished in reversible circuits by adding extra gates. The smallest possible number of reversible logic gates should be used when designing reversible circuits. There are several factors that can affect a circuit's complexity and performance from the perspective of reversible circuit design [27].

1.6.3.2 Basic Reversible Logic Gates

Several significant basic reversible logic gates include:

I.   Feynman Gate

Feynman gate is also known as controlled not gate. This gate has two inputs (A, B) as well as two outputs (X, Y). If inputs are A and B then the output are given by $X = A$ and $Y = A$ XOR B. This gate has a unity quantum cost. A signal can be copied using this gate. Since fan-out is prohibited in reversible logic circuits, a signal copying device called a Feynman gate is employed in its place. The block and logic diagram of Feynman gate is shown Figure 1.15. Table 1.1 provides the Feynman gate truth table.

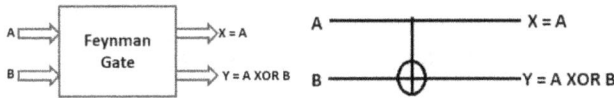

FIGURE 1.15 Feynman gate.

TABLE 1.1
Truth table of Feynman gate

Inputs		Outputs	
A	B	X	Y
0	0	0	0
0	1	0	1
1	0	1	1
1	1	1	0

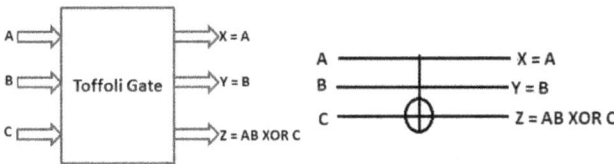

FIGURE 1.16 Toffoli gate.

II. Toffoli Gate

Toffoli gate is a three qubit gate. It has two controls and one target. This gate has three inputs (A, B, C) as well as three outputs (X, Y, Z). If inputs are A, B and C then the outputs are given by X = A, Y = B and Z = (A AND B) XOR C. This gate has 5 quantum cost. The block and logic diagram of Toffoli gate is shown Figure 1.16. Table 1. 2 provides the Toffoli gate truth table.

III. Fredkin Gate

Fredkin gate is a three bit gate. If first bit is only 1 then this gate switch the last two bits. This gate has three inputs (A, B, C) as well as three outputs (X, Y, Z). If inputs are A, B and C then the output are given by X = A, Y = (A'B) XOR (AC) and Z = (A'C) XOR (AB). This gate has 5 quantum cost. The block and logic diagram of Fredkin gate is shown Figure 1.17. Table 1.3 provides the Fredkin gate truth table.

IV. Peres Gate

Peres gate is a three bit gate. This gate has three inputs (A, B, C) as well as three outputs (X, Y, Z). If inputs are A, B and C then the output are given by X = A, Y = A

TABLE 1.2
Truth table of Toffoli gate

INPUTS			OUTPUTS		
A	B	C	X	Y	Z
0	0	0	0	0	0
0	0	1	0	0	1
0	1	0	0	1	0
0	1	1	0	1	1
1	0	0	1	0	0
1	0	1	1	0	1
1	1	0	1	1	1
1	1	1	1	1	0

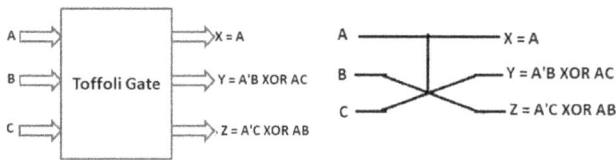

FIGURE 1.17　Fredkin gate.

TABLE 1.3
Truth table of Fredkin gate

INPUTS			OUTPUTS		
A	B	C	X	Y	Z
0	0	0	0	0	0
0	0	1	0	0	1
0	1	0	0	1	0
0	1	1	0	1	1
1	0	0	1	0	0
1	0	1	1	1	0
1	1	0	1	0	1
1	1	1	1	1	1

XOR B and $Z = (AB)$ XOR C. This gate has 4 quantum cost. The block and logic diagram of Peres gate is shown Figure 1.18. Table 1.4 provides the Peres gate truth table.

1.7　COMPARISON BETWEEN SOME POWER REDUCTION TECHNIQUES

The Comparison study between some power reduction techniques in the form of its advantages and disadvantages are given in Table 1.5.

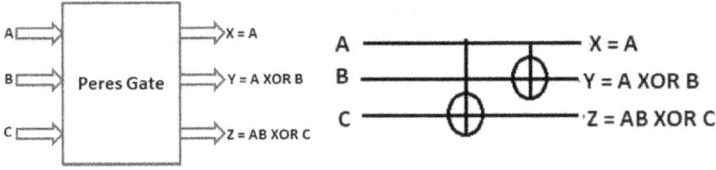

FIGURE 1.18 Peres gate.

TABLE 1.4
Truth table of Peres gate

INPUTS			OUTPUTS		
A	B	C	X	Y	Z
0	0	0	0	0	0
0	0	1	0	0	1
0	1	0	0	1	0
0	1	1	0	1	1
1	0	0	1	1	0
1	0	1	1	1	1
1	1	0	1	0	1
1	1	1	1	0	0

TABLE 1.5
Advantages and disadvantages of power reduction methods

Technique	Advantages	Disadvantages
MTCMOS [18]	Power efficient with zero effect on speed	Required large area
Transistor Stacking [19]	Easy to fabricate and implement	Increase the propagation delay
Sleepy stack[20]	Single threshold transistor	Required large area and control circuit
Clock gating [9]	Medium power saver	Complexity in implementation
Dual V_{DD}[14]	Medium power saver	Required control circuit
DVFS [13]	Good power saver	Complexity in implementation
Adiabatic [22]	Good power saver	Required large area

1.8 SUMMARY OF THE CHAPTER

There are various market segments that are driving the demand for lower power systems. Unfortunately, designing for low power adds another layer of the already challenging design problem, necessitating power as well as performance and area optimization. This chapter discussed various power-optimization methods for CMOS circuits at different levels of design, basic idea of quantum logic and reversible logic circuits. The chip's static power dissipation is rising day by day as a result of the continuous integration. Currently, high speed clock signals contribute far more dynamic power dissipation than chip-based data processing. The world will actually be ruled by QCA technology in the near future.

REFERENCES

[1] Lin, I. C., Lin, C. H., & Li, K. H., "Leakage and Aging Optimization Using Transmission Gate-Based Technique", IEEE Transactions on Computer aided design of integrated circuits and systems, Jan 2013, 32.

[2] Kumar, D. & Tiwari, N., "VLSI Designs for Low Power Applications", International Journal of Engineering Science & Advanced Research, March 2015, 1, 71–75.

[3] Gaur, H. M., Singh, A. K., & Ghanekara, U., "A Review on Online Testability for Reversible Logic", ICECCS, 2015, 384–391.

[4] Roy, K., et al., "Leakage Current Mechanisms and Leakage Reduction Techniques in Deep Sub-micrometer CMOS Circuits". Proceedings of the IEEE, February 2003, 91(2), 305–327.

[5] Tong, Q., Choi, K., & Cho, J. D., "A Review on System Level Low Power Techniques". Proceedings of international SoC Design Conference, 2014.

[6] Sharma, C., "Low Power at Different Levels of VLSI Design and Clock Distribution Schemes", International Journal of Computer Applications in Technology, 2011, 2(1), 88–93. ISSN: 2229–6093.

[7] Bab, A., "Power Optimization Techniques at Circuit and Device Level in Digital CMOS VLSI–A Review", International Journal of Engineering Research & Technology, Nov 2014, 3(11), 375–379.

[8] Gupta, S. & Padave, S., "Power Optimization for Low Power VLSI Circuits", International Journal of Advanced Research in Computer Science and Software Engineering, Mar 2016, 6(3), 96–99.

[9] Kathuria, J., Ayoubkhan, M., & Noor, A., "A Review of Clock Gating Techniques", MIT International Journal of Electronics and Communication Engineering, Aug 2011, 1(2), 106–114.

[10] Usami, K., & Horowitz, M., "Clustered Voltage Scaling Technique for Low Power Design", Proceedings of the 1995 International Symposium.

[11] Dokić, B. L., "A Review on Energy Efficient CMOS Digital Logic", Engineering, Technology & Applied Science Research, 2013, 3, 552–561.

[12] Sivakumar, R. & Jothi, D., "Recent Trends in Low Power VLSI Design", International Journal of Computer and Electrical Engineering, Dec 2014, 6(6), 509–523.

[13] Pillai, P., & Shin, K. G., "Real-Time Dynamic Voltage Scaling for Low-power Embedded Operating Systems", Proceedings of the Eighteenth ACM Symposium on Operating Systems Principles, Dec 2011.

[14] Gayasen, A., Lee, K., Vijay Krishnan, N., Kandemir, M., Irwin, J., & Tuan, T., "A Dual-VDD Low Power FPGA Architecture", In: Becker, J., Platzner, M., Vernalde, S. (eds) Field Programmable Logic and Application. FPL 2004. Lecture Notes in Computer Science, vol. 3203. Springer, Berlin, Heidelberg. July 2004.

[15] Kang, "CMOS Digital Integrated Circuits", 3rd Edition, McGraw Hill.

[16] Texas Instruments. Adaptive voltage scaling technology up to 60% energy savings for digital core operation.

[17] Hartman, M., "Processor Energy Savings Through Adaptive Voltage Scaling", Issue of Portable Design, March 2008.

[18] Sreenivasulu, P., Khadar Khan, P. Dr. Srinivasa Rao, K., & Dr. Vinaya Babu, A. "Power Scaling in CMOS Circuits by Dual Threshold Voltage Technique", International Journal of Engineering and Innovative Technology (IJEIT), July 2013, 3(1), 128–133.

[19] Narendra, S., Borkar, S. De, V. Antoniadis & Chandrakasan A. "Scaling of Stack Effect and Its Application for Leakage Reduction", Proceedings of the International Symposium on Low Power Electronics and Design, Aug 2001, 195–200.

[20] Park, J. C. & Mooney, V. J., "Sleepy Stack Reduction of Leakage Power", IEEE Transactions on Very Large-Scale Integration (VLSI) Systems, Nov 2006, 14(11), 1250–1263.

[21] Bhuvana, B. P., Manohar, B. R., & Kanchana Baskaran, V. S., "Adiabatic Logic Circuits Using FinFETs and CMOS–A Review", International Journal of Engineering and Technology, 2016, 8(2), 1256–1270.

[22] Sunil Gavaskar Reddy, Y. & Rajendra Prasad, V. V. G. S., "Power Comparison of CMOS and Adiabatic Full Adder Circuits", International Journal of Scientific & Engineering Research, 2011, 2(9), 1–5.

[23] Lent, C. S., Tougaw, P. D., Porod, W., & Bernstein, G. H. "Quantum Cellular Automata", Nanotechnology, 1993, 4, 49–57.

[24] Shivkumar, R. & Jothi, D, "Recent Trends in Low Power VLSI Design", International Journal of Computer and Electrical Engineering, Jan 2014, 6(6), 509–523.

[25] Kim, Y. B. "Challenges for Nanoscale MOSFETs and Emerging Nanoelectronics", Transaction on Electrical and Electronic Materials, 2010, 11, 93–105.

[26] Kumari, P., & Gurumurthy, K. S. "Quantum Dot Cellular Automata: A Review", Proceedings of the IEEE International Conference on Advances in Computing and Information Technology, 2014, 3(1), 51–55.

[27] Gaur, H. M. & Singh, A. K. "Design of Reversible Circuits with High Testability", Electronics Letters, 23 June 2016, 52(13), 1102–1104.

[28] Song, Z., Xie, G., Cheng, X., Wang, L. & Zhang, Y. "An Ultra-Low Cost Multilayer RAM in Quantum-Dot Cellular Automata", IEEE Transactions on Circuits and Systems II: Express Briefs, 2020, 67(12), 3397–3401.

2 Investigations on Designing of Adders, Multiplexers and Flip-Flops for Fast Memories Development in QCA Technology

*Paramjeet, Saptarshi Gupta and
Hari Mohan Gaur*

2.1 INTRODUCTION

CMOS technology has been the dominant technology since the mid of 20th century. Gordon Moore, in 1965, predicted that number of transistors in an integrated circuit will be doubled every 18 months [5]. This prediction is well known as Moore's law, which has been followed by the electronics industry for more than forty years. But, now it seems that Moore's law will not be able to direct the electronics industry through VLSI technology as the CMOS transistors start to face the physical limits [6]. Further miniaturization results in other unavoidable issues which affect the performance of the device or circuit. This technology has now a number of problems at nanoscales [7, 8, 9]. Some of the issues arising due to miniaturization include (1) high power dissipation/ consumption, (2) quantum mechanical side effects, (3) leakage currents, (4) low fan-out, and (5) oxide thickness and thermal reliability.

Further, size of transistor is also moving towards quantum limits. As we move towards further miniaturization, some of the issues viz. leakage currents, tunneling, heat dissipation, quantum mechanical effects, and interconnect dominance, etc. affect the performance of CMOS circuits [1]. Thus, CMOS technology will not be able to drive the electronics industry in near future due to its limitations. Hence, either the logic or the technology needs to be changed. It is a great challenge to find out a technology that can be thought as a replacement to CMOS technology. Researchers are working on some emerging technologies viz. Quantum-Dot Cellular Automata, Molecular Electronics, Spin Wave Devices, Single Electron Transistor, Silicon Nanowires, etc. Amongst these alternatives, Quantum-Dot Cellular Automata has drawn the attention of the researchers for designing the circuits at nanoscale with ultra-low power dissipation, high performance and reduced feature size [2]. QCA

DOI: 10.1201/9781003361633-2

with reversible logic will produce better results in terms of power consumption [3, 4].

The circuits designed through QCA technology have been found to have high speed, low power and consume minimum area. The combinational logic circuits like multiplexers, de-multiplexers, adders, comparators, encoders/decoders and other structures have been designed by various researchers. The multiplexers/ de-multiplexers and adders play a vital role in designing of complex circuits like memories, ALU, etc. This paper analyses various adders, multiplexers/ de-multiplexers, flip-flops. Efficient RAM cell analysis is produced with this background that were introduced by researchers from the literature.

The chapter is organized as follows: Section 2.2 presents a brief on the QCA fundamentals, designing approaches and implementation techniques. Section 2.3 provides reviews and investigations on the designing of adders followed by multiplexers and de-multiplexers in Section 2.4. For the beginning of memory designing, the building blocks are flip-flops which are investigated in Section 2.5 and finally, the core part of the paper that explains the memory designing survey which is provided in Section 2.6 followed by the summary of the chapter in Section 2.7.

2.2   QUANTUM-DOT CELLULAR AUTOMATA: A FUTURE TECHNOLOGY

Quantum-Dot Cellular Automata is nanoscale technology which was invented by C.S. Lent in 1993 [10]. QCA technology has the attractive features that make it better than CMOS technology. This technology allows the designing of the circuit at the nanoscale. High density, low power and high speed devices can be produced by this technology. In this technology, logical states or values are represented by position of electrons instead of voltage levels.

2.2.1   QCA Cell

The basic unit in the QCA technology is QCA Cell consisting of four quantum dots as depicted in Figure 2.1(a).

Open circles represent the quantum dots which are connected by tunnel junctions [11]. Out of these four quantum dots, two are occupied by electrons represented by solid dots. These two electrons can move to neighboring dots of the cell, but due to coulombic repulsions between these electrons, they tend to occupy antipodal dots of the cell to maintain the longest distance between them as shown in the Figure 2.1(b). It can be concluded that these two electrons will never stay at the dots of the cell on the same edge. Thus, there can be two arrangements for the electrons to occupy the stable states. Depending on the positions of electrons, there can be two stable polarizations and accordingly, these two polarizations represent binary one or zero as illustrated in the Figure 2.1(b).

When two cells are placed in vicinity of each other, there will be coulombic interaction between the electrons of both cells, i.e. electrons of one cell will interact with electrons of other cell and in this way one cell can affect the polarization of another cell. This can be understood through the Figure 2.2, where cell 1 determines the state of cell 2.

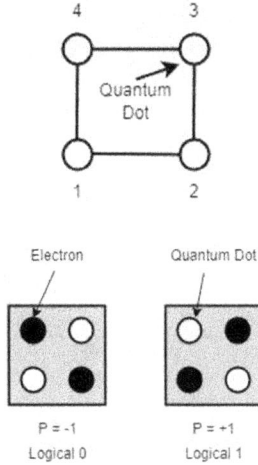

FIGURE 2.1   QCA cell.

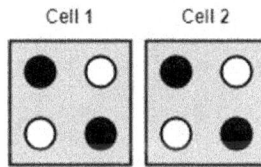

FIGURE 2.2   QCA cells coupling.

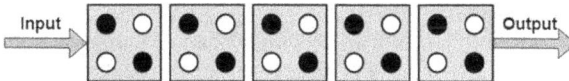

FIGURE 2.3   90° binary wire.

It is clear from Figure 2.2 that one polarized cell is able to polarize the other neighboring cell to the same polarization. Thus, the cells exchange information through coulombic interaction among the cells.

2.2.2   QCA WIRES

QCA binary wires are different from conventional wires in terms of design, structure and functioning. In fact, no current flows in QCA wires. Further, these wires are designed through coupling of QCA cells and coulombic interaction between the neighboring cells will propagate the information. Depending on the way of arranging the cells, there are two designs of QCA cell. These are called as 90° binary wire and 45° binary wire.

90° Binary Wire: This is the simplest wire. The structure of 90° wire is shown in Figure 2.3. The QCA cells are arranged horizontally or vertically.

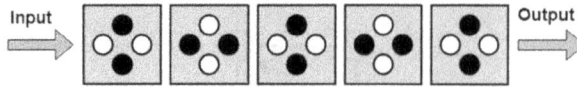

FIGURE 2.4   45° binary wire.

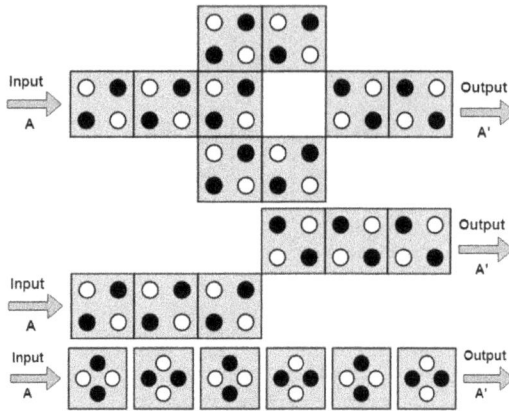

FIGURE 2.5   Inverter gate.

When an input is applied to the first cell, it interacts with the second cell and polarizes it. Now, the second cell polarizes the third cell and in this way the information is propagated from the first cell to the last cell.

45° Binary Wire: The second type of wire is 45° binary wire. In this wire, the cells are first rotated at 45° to their centers and then arranged horizontally or vertically as illustrated in Figure 2.4. It is clear from the figure that two consecutive cells have opposite polarization.

In this way, both logic '0' and logic '1' propagate simultaneously and thus, there is no need of inversion circuitry for getting the inverted value.

2.2.3   BASIC QCA GATES

Coupling of QCA cells through coulombic interaction is the basic concept which is used to design the various logic functions. Logic gates like QCA inverter, majority gate and other basic gates can be designed with the proper arrangements or combination of cells.

2.2.3.1   QCA Inverter Gate

Inverter/ not gate is a very important and basic element in digital circuits. Various types of inverters have been designed using QCA technology. Out of various designs, three basic designs for the inverter are illustrated in Figure 2.5. The inverter design should be properly selected as the density as well as reliability depends on the design.

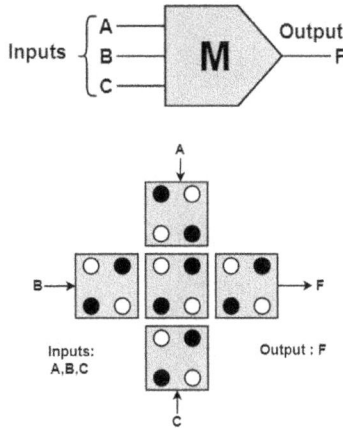

FIGURE 2.6   Majority gate.

As shown in the Figure 2.5(a), the input signal is propagated through two different paths and then the input signal is inverted. This design is more reliable. Figure 2.5(b) shows the other design for the inverter. In this design the output cell is placed at 45° above or below the input cell. This is called as 45° inverter. One more design of inverter is depicted in Figure 2.5(c) which utilizes the 45° wire with an even number of cells.

2.2.3.2   QCA Majority Gate

It is the main gate which is used to design the various logic circuits. The symbol and logic function for this gate is shown in Figure 2.6(a) and Figure 2.6(b), respectively.

As is clear from Figure 2.6, this gate consists of five QCA cells, out of which, three cells are for inputs and one for output. The fifth cell is called as device cell located in the center. The gate is called as majority gate as the output goes to the same state which at least two out of three input cells possess, i.e. output goes towards the majority votes of input cells. Hence, this gate has been named as majority gate.

The logical equation of the majority gate can be represented as shown below:

$$MV\ (A, B, C) = AB + BC + AC$$

Majority gate can be used as AND gate and OR gate by setting one input to logic '0' or '1', respectively, as shown in Figure 2.7.

2.2.4   QCA Clocking

Clock signal is used to synchronize and control the propagation of information in QCA circuit [12]. Clock signal is the only power supply to the QCA circuit, i.e. no other power supply is required for the operation of the QCA circuits. QCA circuit is divided into four areas called clocking zones. These zones are called Switch, Hold, Release, and Relax phase. Each phase is 90° out of phase to the preceding phase. The four phases of the clock signal are explained below with help of the Figure 2.8.

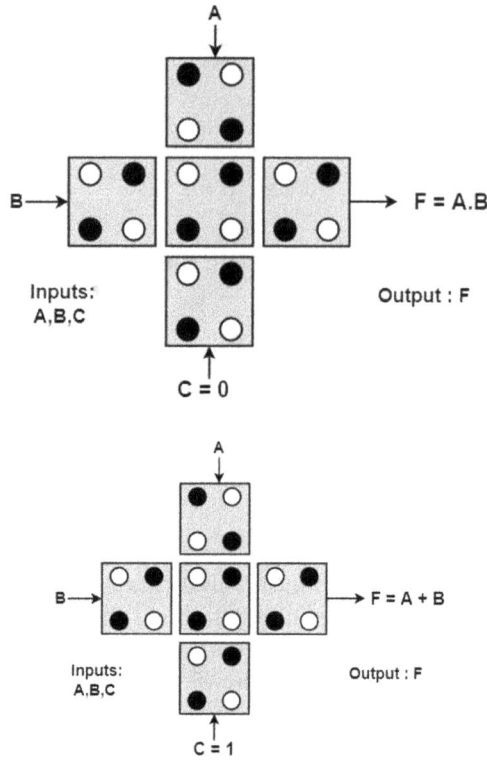

FIGURE 2.7   (a) Majority gate as AND gate (b): Majority gate as OR gate.

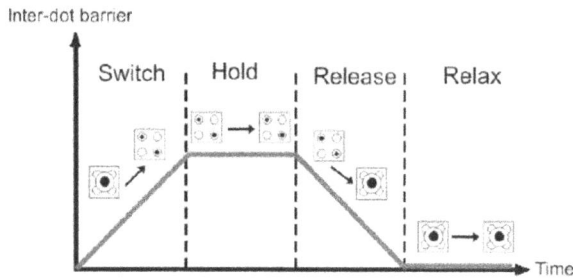

FIGURE 2.8   QCA clocking.

- **Switch Phase:** When the clock moves from low to high, the phase is called switch phase. As clear from the figure, the clock signal rises from low state to high state in adiabatic form. The rise in clock signal raises the inter-dot barrier, which allows the polarization of the cell due to the influence of the neighboring cells. In this way; the computations are done under switch phase.
- **Hold Phase:** As the name suggest, the polarization of the cell is held in this phase with the help of holding the inter-dot barrier at high. Thus, the electrons stay at their existing positions and will not be able to travel to the quantum

dots in vicinity to them. Thus, the cell is able to hold the polarization during this phase.

- **Release Phase:** Under this phase, the clock moves from high state to the low state and thus the inter-dot barrier goes down towards zero, which makes the cell un-polarized.
- **Relax Phase:** During this phase the clock is low and the cell remains un-polarized.

2.2.5   QCA IMPLEMENTATION TECHNIQUES

There are various physical implementation techniques for QCA circuits, keeping the basic concept of bi-stable cell that can interact with its neighboring cells. The main four implementation techniques are described below [13]:

2.2.5.1   Metal-Island QCA

The metal-island QCA cell consists of metal islands made of aluminum instead of coulombically coupled quantum dots. Capacitive coupled metal islands are used to construct a cell [14]. The metal islands, called dots, are connected with aluminum oxide tunnel junctions for the tunneling of electrons between the dots. Metal-island QCA works on very low temperature, thus, it is very difficult to fabricate the complex circuits which can operate at room temperature. This is the main obstacle to accept this technique for the fabrication of QCA circuits in future.

2.2.5.2   Semiconductor QCA

Semiconductor material such as InAs/ GaAs and GaAs/ AlGaAs is used to fabricate quantum dots [15]. QCA cells can be fabricated with the same advanced CMOS technology using the semiconductor material. Cell polarization depends on the position of charge and computation happens based on the interaction or coupling of a cell with its neighboring cells. The main issue with this technique is that mass production of QCA cell at nanoscale with the existing CMOS technology is not feasible.

2.2.5.3   Molecular QCA

Molecular technique has drawn a lot of attention of researchers as this technique is able to work at room temperature. The cell is formed with a single molecule. The circuits implemented with this technique possess high density and high switching speed and mass production of devices is also possible with Molecular QCA technique [16].

2.2.5.4   Magnetic QCA

Magnetic implementation of Quantum-Dot Cellular Automata (MQCA) was proposed by Cowburn and Welland [17]. A nanomagnet consisting of a single circular nanodot is used as QCA cell. The nanodots which are made of magnetic supermalloy are placed in a straight line fashion and an oscillating field is applied to the dot. Magnetic QCA cell can operate at room temperature but the operating frequency is low.

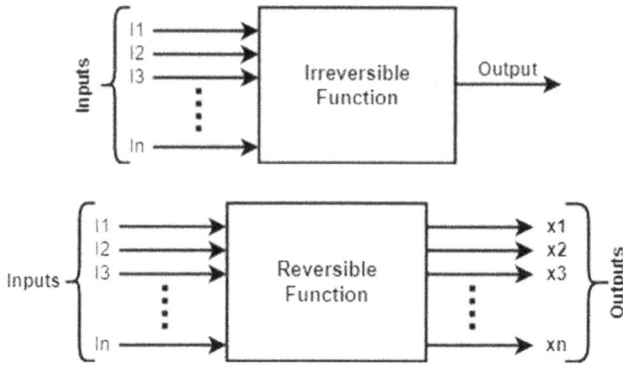

FIGURE 2.9   Irreversible and reversible functions.

2.2.6   LOGIC DEVELOPMENT: IRREVERSIBLE / TRADITIONAL AND REVERSIBLE LOGIC

Logic functions can be designed with two approaches, i.e. irreversible and reversible logic. As per R. Landauer information is lost in irreversible computation and an amount of kTln2 Joule of heat energy is dissipated per bit information loss [3]. Thus, this amount increases with the number of bits lost. The conventional circuits use the irreversible gates in which the heat energy is dissipated because of information loss during the computation.

In 1973, C.H. Bennett [4] proposed the reversible computation concept for resolving the issue of information loss. If a logic function produces a distinct output state for each input state, it will be defined as a reversible logic function [18]. In a reversible system, the inputs can be reproduced from the outputs. In this way, ideally, no information will be lost as the same can be recovered. The metrics such as number of gates, wires, garbage, and stages of quantum, etc. decides the performance of the design [19, 49].

As shown in the Figure 2.9(b), reversible logic uses '**n × n**' reversible functions. These functions possess bijective mapping between the inputs and outputs.

Reversible logic and QCA can be combined to develop the combinational and sequential circuits. The circuits designed and developed with these two paradigms present better performance in terms of power and speed. A lot of work is being done by the researchers to develop the efficient circuits. Thus, the reversible logic implemented through QCA technology can be the suitable alternative of CMOS technology.

2.3   DESIGNING OF ADDER

Various researchers have designed a number of combinational as well as sequential circuits using QCA technology. All the researchers and authors have tried to design the circuits with minimum complexity, area, and clock latency. An electronic circuit whose output is dependent on the present combination of inputs only, is called as combinational circuit. The main combinational circuits are adders, multiplexers, de-multiplexers, etc.

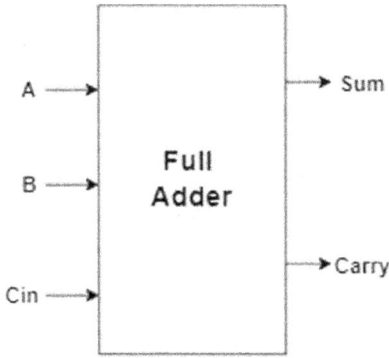

Inputs			Outputs	
A	B	C_{in}	Sum	Carry
0	0	0	0	0
0	0	1	1	0
0	1	0	1	0
0	1	1	0	1
1	0	0	1	0
1	0	1	0	1
1	1	0	0	1
1	1	1	1	1

FIGURE 2.10   Full adder block diagram and truth table.

Full Adder is a digital combinational circuit that adds three bits, i.e. two inputs and a carry generated from previous stage. The generated outputs Sum and Carry are represented by the following Boolean expression.

$$Sum = A \text{ XOR } B \text{ XOR } C_{in}$$

$$Carry = A. \ B + B. \ C_{in} + A. \ C_{in}$$

The block diagram and truth table of Full Adder is shown in Figure 2.10.

Addition is considered as the most important mathematical operation because adder is mostly used to implement the other operations like subtraction, multiplication and division. Thus, there is great demand of efficient adders for the designing of the various circuits. Many researchers have designed a number of adders in QCA technology. A lot of work has been done by the researchers for the designing of energy and area efficient adder circuits with minimum delay. The various authors have proposed the different structures following the coplanar or multi-layer approach. Further, the authors also designed the coplanar adder with the help of either 3-input MV or 5-input MV or XOR Gate or with a combination of these basic elements.

P. D. Tougaw et al. in 1994 proposed the first QCA full adder with the use of five 3-input majority gates and three inverters. The design employed 192 QCA cells in single layer [50]. By **Angizi et al.** in 2014, one 3-input and one 5-input majority gate with one inverter and two crossovers were utilized to build full adder design. This consumes 95 cells in an area of 0.09 μm^2 and 05 clock zones to produce the required output [52]. **Hashemi et al.** in 2015 proposed a compact design of full adder that utilized only 71 QCA cells consuming an area of 0.06 μm^2. The proposed adder design has latency of 1.25 which increases the cost function of the circuit. The circuit has been designed with the use of five-input majority gate and a robust coplanar crossover approach [53]. **Sasamal et al.** in 2018 proposed a design of full adder with the use of one 3-input majority gate, one 5-input majority gate and two inverters. The complexity of the circuit is 46 QCA cells which consume an area of 0.035 μm^2. The clock latency of the circuit is 1.0 [55]. **Girija et al.** in 2019 presented a full adder with the use of one 5-input majority gate and one 3-input majority gate. The circuit uses 38 QCA cells and consumes an area of 0.04 μm^2. The author claims that the circuit offers no clocking delay [57]. **Saeid Zoka et al.** in 2019 proposed a full adder including one 3-input majority gate, two 5-input majority gates and six number of inverters. The architecture has a complexity of 83 QCA cells consuming an area of 0.09 μm^2. The clock latency of the circuit is 1.5 [59].

Further, **Kianpour et al.** in 2014 presented a full adder with the use of three 3-input majority gates and two inverters. The architecture is designed with 69 QCA cells in single layer and area consumed by the circuit is 0.07 μm^2. The latency of the circuit is 0.75 [51]. **Abedi et al.** in 2015 designed a full adder circuit which utilizes three 3-input majority gates, two inverters and a crossover. The architecture is designed with the use of 59 QCA cells consuming an area of 0.043 μm^2. The latency of the circuit is 1.0. The circuit is designed in a single layer [54]. **Z. Altarawneh et al.** in 2021 presented an efficient full adder with the use of 14 QCA cells and consuming an area of 0.012 μm^2. The proposed design has a clock latency of 0.05 [61]. The proposed structure has been designed with the use of a 3-input majority voter gate and a 3-input XOR gate proposed by R. Laajimi in 2018 [62].

Also, the adder can be designed without the use of MVs. In this approach, the authors designed the circuits with the use of XOR gate and some researchers also proposed structures on the basis of cell interaction. **Ismail G. et al.** in 2021 realized an adder including only 19 QCA cells consuming an area of 0.01μm^2. The clock latency of the circuit is 0.5. Firstly, the author proposed an optimized XOR gate and then, with the help of this proposed XOR gate, the full adder is designed [60]. An adder utilizing only 15 cells and area of 0.007 μm^2 was proposed by **Ali Majeed and Esam Alkaldy** in 2022 using cell interaction based approach. The circuit has a clock latency of 0.5 [64]. The literature survey done for the coplanar adder is summarized in the Table 2.1.

Limited work has been proposed in designing of circuits using multi-layer approach. A few of the full adder architectures with the multi-layer approach are detailed as follows. **Sen et al.** in 2013 presented a full adder constructed with the help of one 3-input majority gate and one 5-input majority gate. The complexity of the circuit is 31, i.e. it consists of 31 QCA cells and consumes an area of 0.01 μm^2. The adder has been found to have the latency of 0.50 [56]. **Mostafa Sadeghi et al.** in

TABLE 2.1
Summary of literature survey for coplanar full adder

Full Adder Structure	No of Majority Gate		No. of Inverters Used	No. of Crossover	Area (In μm²)	Complexity (No. of QCA Cells)	Latency (Clock Cycle)	Circuit Cost = Area * Cells * Latency
	3-Input MV	5-Input MV						
[51], 1994	5	0	3			192		
[52], 2014	3	0	2		0.07	69	1	4.83
[53], 2014	One 3-input MV and one XOR Gate	0	1	2	0.09	95	1.25	10.69
[54], 2015	1	1			0.06	71	1.25	5.36
[55], 2015	3	0	2	1	0.043	59	1	2.54
[56], 2018	1	1	2		0.035	46	1	2.59
[58], 2019	1	1			0.04	38	0	0
[60], 2019	1	2	6		0.06	46	1.25	3.45
[61], 2021	0	0	0	0	0.01	19	0.5	0.10
[62], 2021	1	0	0	0	0.012	14	0.5	0.10
[65], 2022	0	0	0	0	0.007	15	0.5	0.053

TABLE 2.2
Summary of literature survey for multi-layer full adder

Full Adder Structure	No of Majority Gate		No. of Inverters Used	Area (In μm²)	Complexity (No. of QCA Cells)	Latency (Clock Cycle)	Circuit Cost = Area * Cells * Latency
	3-Input MV	5-Input MV					
[57], 2013	1	1		0.01	31	0.5	0.16
[59], 2016	1	1	2	0.0598	35	1.25	2.62
[64], 2020	1	1	1	0.01	30	0.75	0.23

2016 devised a full adder circuit using one 3-input majority gate, one 5-input majority gate and two inverters. The circuit has been designed in layers consuming an area of 0.0598 μm² which consist of 35 QCA cells. The clock latency of the proposed adder is 1.25 [58]. **Lei Wang and Jie Yan** in 2020 proposed a full adder circuit with the use of one inverter, one 3-input majority gate and one 5-input majority voter gate. The circuit makes the use of 30 QCA cells consuming an area of 0.01 μm². The circuit has a clock latency of 0.75 [63]. The summary of literature survey for multi-layer full adder is given in Table 2.2.

2.4 DESIGNING OF MULTIPLEXER AND DE-MULTIPLEXER

As there is no feedback ability in quantum circuits and obviously in QCA circuits as well, it can be done by providing delay in the path during the flow of information. Multiplexer and de-multiplexers are the basic and important building block for the nanocomputing sequential designs. The review and investigations for these logic circuits are provided in this section.

2.4.1 MULTIPLEXER DESIGNS

A multiplexer being a digital combinational circuit is used to transmit the signal from one of many inputs to single output depending on the status of the control inputs, i.e. select line. There are 2^N inputs, one output and N Select Line in 2^N:1 multiplexer. The block diagram of 2^N:1 multiplexer and 2:1 multiplexer along with truth table is given in Figure 2.11.

The Boolean expression for 2:1 multiplexer is Y= S'.A + S.B. Thus, when select line, i.e. S=0, then input A will be transmitted to output line Y and if S=1, then output will be equal to B. In the same way, the signal values from one of inputs will be transferred to output depending upon the value of the select lines in case of higher order multiplexer viz. 4:1, 8:1, 16:1, and so on. The QCA multiplexer was addressed for the first time in a primitive work by Gin et al. in 1999 [92]. Various researchers have devised different multiplexer architectures following different approaches in QCA technology. The researchers have made efforts to produce efficient multiplexer structure in terms of complexity, area, clock latency, and power consumption, etc. and

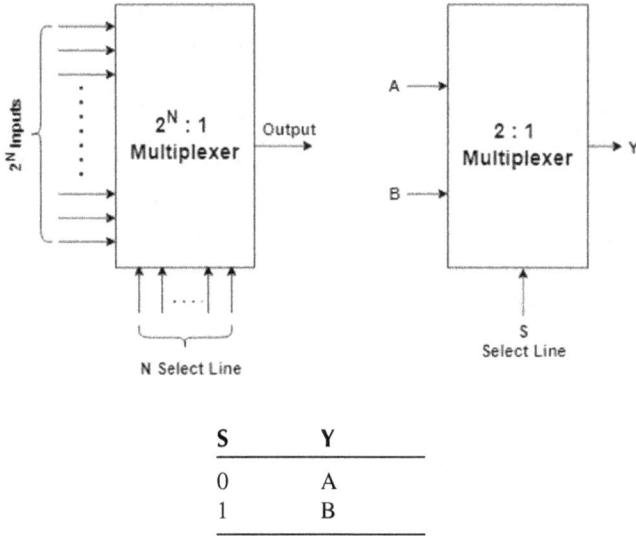

FIGURE 2.11   (a) 2^N: 1 multiplexer block diagram (b) 2:1 multiplexer block diagram (c) truth table.

adopted different methodologies for designing such as coplanar/ multi-layer, majority voter/ XOR gate/ cell interaction based. The circuits can be designed with the help of either 5-inputs majority gates or 3-input majority gates or using both gates simultaneously. The detail of some of such architectures is summarized below.

Bibhash Sen et al. (2012) devised a novel 2:1 multiplexer with the use of majority gate. The architecture consists of 19 QCA cells consuming an area of 0.02 μm² [34]. The clock latency of the designed circuit is 0.75. The presented robust multiplexer has been used in designing of ALU that performs ten different basic arithmetic-logic functions. **M. Kianpour et al.** (2013) proposed a 2 to 1 multiplexer with the use of 22 QCA cells that cover an area of 0.03 μm² and latency of the circuit was found to be 0.75. The higher order multiplexers were also suggested by the author through the use of proposed 2:1 multiplexer. The architecture has been designed with the help of three majority gates and one inverter [37]. **Sabbaghi-Nadooshan et al.** (2013) proposed a 2:1 multiplexer circuit with the use of three 3-input majority gates and one inverter. The complexity of the circuit is 26 QCA cells which consume an area of 0.02 μm². The architecture has a clock latency of 0.5 [40]. **B. Sen et al.** (2015) proposed a modular and efficient 2:1 multiplexer with the use of only 23 QCA cells and latency of 0.5. The area requirement of proposed 2:1 multiplexer was 0.02 μm² [20]. The proposed 2:1 multiplexer was utilized to design a CLB in QCA. A modular approach in developing the multiplexer of higher order through modular approach with cascading of lower order multiplexers has been discussed. **Rashidi et al.** (2016) designed an architecture of 2:1 multiplexer consisting of two majority gates, a rotate majority gate and an inverter gate. It was designed with the help of 15 QCA cells with an area of 0.01 μm² and latency of the circuit is 0.5 [21].

Further, a 4:1 multiplexer having 107 QCA cells and an area of 0.15 μm^2 was also proposed by the authors. Moreover, 8:1 multiplexer consisting of 293 QCA cells and consuming an area of 0.58 μm^2 was constructed. **J. C. Das et al.** (2016) proposed a 2:1 multiplexer circuit that requires three majority voter gates, one inverter and two clock zones, i.e. 0.5 clock latency [22]. The circuit contains 17 cells and consumes 0.012 μm^2 area. The area utilization was only 47.22%, which requires improvement. **Jin-Seong et al.** (2016) constructed a 2:1 multiplexer using NAND gate with the help of majority gate and inverters. The proposed design utilized three majority gates and three inverters. The architecture consists of 31 QCA cells consuming an area of around 0.01 μm^2 with a clock latency of 0.5 [35]. **Rashidi et al.** (2017) designed a 2:1 multiplexer architecture with two rotated majority gates (RMG), one original majority gate (OMG) and one inverter gate. The circuit is implemented with QCA Designer through the bi-stable implementation engine, as bi-stable implementation engine implements the design faster. The area requirement of the designed 2:1 multiplexer is found to be 0.02 μm^2 and complexity of the circuit is 17 QCA cells. The latency of the circuit is 0.5. Further, the authors also proposed the higher order multiplexers, i.e. 4:1 and 8:1 multiplexers with improved efficiency in terms of complexity, area, and latency [23].

Khosroshahy et al. (2017) proposed a unique method to reduce the number of external input cells used to provide fixed input in a circuit. Three approaches for three different situations have been discussed. One for situation where only single type of fixed value (0 or 1) is required, second for the situation for which two types of fixed value is required but the number of such values is same. The third approach is for the situations where two types of fixed value is required but the number of such values is not the same. The authors applied the approach to multiplexers and found the improved results as compared to the previous designs. The proposed 2:1 multiplexer was designed with 36 numbers of QCA cells and latency of 1 [25].

F. Ahmad (2017) suggested a method for the designing of 2:1 multiplexer and the proposed multiplexer has been found to have 16 QCA cells that occupy an area of 0.01 μm^2. The latency of proposed circuit is 0.5. The design uses two rotated majority gates (RMG), one original majority gate (OMG) and one inverter gate [26]. **Ahmadpour et al.** (2018) proposed a fault-tolerant 2:1 multiplexer that shows the good tolerance for single and double-cell omission defects in the circuit. The multiplexer has been designed with the help of proposed fault tolerant three inputs majority gate. The complexity and latency of the multiplexer is found to be 35 QCA cells and 0.25, respectively. The design is found more robust against variations in the temperature [27]. **M. Mosleh** (2018) proposed a 2:1 multiplexer with the help of new designed majority gate (MV32) having three inputs and two outputs with 11 QCA cells. The multiplexer circuit has been designed with 21 QCA cells that occupy an area of around 0.01 μm^2. The circuit has been found to have a latency of 0.75 that make it slower as compared to other equivalent circuits [28].

Sohankumar Dahana et al. (2018) presented a 2:1 multiplexer with complexity of 24 QCA cells consuming area of 0.03 μm^2. The clock latency of the proposed structure is 0.5 and crossover is coplanar [33]. **L Xingjun et al.** (2019) suggested a 2:1 multiplexer circuit with one majority gate and one novel XOR gate. The 2:1 multiplexer has been used for the designing of higher order multiplexers, i.e. 4:1 and 8:1

multiplexers. The proposed 2:1, 4:1 and 8:1 multiplexer circuits consist of 22, 92 and 260 QCA cells and cover 0.03, 0.12, and 0.40 μm^2 of area, respectively. The latency of the proposed 2:1, 4:1 and 8:1 multiplexers are 0.75, 1.25 and 2.25, respectively. The important aspect of these structures is that all the structures have been designed without any crossover [29]. **Jeon J.C.** (2021) presented a 2 to1 multiplexer using QCA technology, complexity of which was 19 QCA cells consuming an area of 0.01 μm^2. The suggested design was having a clock latency of 0.5. The structure has been designed with use of three majority gates for AND operation and four inverters [36]. **Hamid Rashidi et al.** (2021) developed a 2:1 multiplexer with use of 16 QCA cells which consume an area of 0.01 μm^2. The circuit consume 0.5 clock cycle to produce the output. This circuit has been designed with the help of one rotated majority voter gate, two original majority voter gates and one inverter gate. The architecture has been designed in single layer [41]. **Angshuman Khan et al.** (2022) proposed a QCA multiplexer which has a complexity of 28 QCA cells covering an area of 0.05 μm^2. The suggested 2:1 multiplexer has a clock latency of 0.75. The design has been proposed with no wire-crossing. The design has been found to be energy efficient as compared to the equivalent circuits proposed by various researchers in the past [38]. **Mangali Mahesh et al.** (2022) proposed a design of 2:1 multiplexer in which 45° rotated QCA cells are used as selection line. The circuit includes 30 QCA cells which occupy an area of 0.04 μm^2. The latency of the proposed architecture is 0.5. Further, the author also presented an optimized QCA 2:1 multiplexer utilizing 26 cells and the designed circuit can be used to implement various gates by the selection of suitable inputs/ selection line. Higher order multiplexers, i.e. 4:1 and 8:1 multiplexer can also be constructed with use of proposed 2:1 multiplexer [32]. Summary of the literature survey for majority voter based multiplexer designs is detailed in Table 2.3.

Some of the authors have also proposed the multiplexer structures designed on cell based interaction, i.e. without the use of majority gates, which are summarized below.

Asfestani et al. (2017) constructed a QCA 2:1 multiplexer without majority gate, utilizing 12 QCA cells and an area of 0.01 μm^2. The latency of the designed circuit has been claimed to be 0.25 only. No crossover is required in this structure. Also, the author proposed a QCA 4:1 multiplexer consisting of 59 cells on an area of 0.08 μm^2 [24]. **E. AlKaldy et al.** (2020) proposed a 2:1 multiplexer that consists of 11 QCA cells consuming 0.01 μm^2. The latency of the suggested multiplexer is 0.25. The author also proposed a 4:1 multiplexer consisting of 37 QCA cells consuming an area of 0.03 μm^2 with latency of 0.75 [30]. **Amjad Almatrood et al.** (2021) presented two different architectures of 2: 1 multiplexer designs with lower energy dissipation. Both the circuit designs have the same complexity of 13 QCA cells consuming an area of 0.01 μm^2 [31]. The latency of the both the architecture is 0.5. The proposed multiplexer circuits with two clock zones have average energy dissipation of 0.545 meV and 0.479 meV, respectively. **Seyed-Sajad Ahmadpour et al.** (2021) designed a 2:1 multiplexer with a lower complexity of ten QCA cells consuming an area of 0.03 μm^2. The clock latency of this circuit is 0.5 [39].

The summary of the literature survey for cell interaction based multiplexer is detailed in Table 2.4.

TABLE 2.3
Summary of literature survey for majority voter based multiplexer designs

2:1 MUX Structure	No of Majority Gate		No. of Inverters Used	No. of Crossover	Area (In μm²)	Complexity (No. of QCA Cells)	Latency (Clock Cycle)	Single / Multi-Layer	Circuit Cost = Area * Cells * Latency
	3-Input MV	3-Input RMG							
[35], 2012	3	0	1	0	0.02	19	0.75	Single	0.29
[41], 2013	3	0	1	0	0.02	26	0.5	Single	0.26
[38], 2013	3	0	1	0	0.03	22	0.75	Single	0.50
[21], 2015	3	0	1	0	0.02	23	0.50	Single	0.23
[22], 2016	2	1	1	0	0.01	15	0.5	Single	0.08
[23], 2016	3	0	1	0	0.012	17	0.5	Single	0.10
[36], 2016	3	0	3	0	0.01	31	0.5	Single	0.16
[24], 2017	1	2	1	0	0.02	17	0.5	Single	0.17
[26], 2017	1	0	1	0	-	36	1	Single	-
[27], 2017	1	2	1	0	0.01	16	0.5	Single	0.08
[28], 2018	3	0	1	0	-	35	0.25	Single	-
[29], 2018	One three input 2 output MV	0	0	1	0.01	21	0.75	Multi-layer	0.16
[34], 2018	3	0	1	0	0.03	24	0.5	Single	0.36
[30], 2019	3	0	1	0	0.03	22	0.75	Single	0.50
[37], 2021	3	0	4	0	0.01	19	0.5	Single	0.10
[42], 2021	2	1	1	0	0.01	16	0.5	Single	0.08
[33], 2022	3	0	2	0	0.04	30	0.5	Single	0.6
[39], 2022	3	0	1	0	0.05	28	0.75	Single	1.05

TABLE 2.4

Summary of literature survey for cell interaction based multiplexer

2:1 MUX Structure	No of Majority Gate		No. of Inverters Used	No. of Crossover	Area (In µm²)	Complexity (No. of QCA Cells)	Latency (Clock Cycle)	Single / Multi–Layer	Circuit Cost = Area * Cells * Latency
	3-Input MV	3-Input RMG							
[25], 2017	0	0	0	0	0.01	12	0.5	Single	0.06
[31], 2020	0	0	0	0	0.01	11	0.25	Single	0.03
[40], 2021	0	0	0	0	0.03	10	0.5	Single	0.15
[32], 2021	0	0	0	0	0.01	13	0.5	Single	0.07

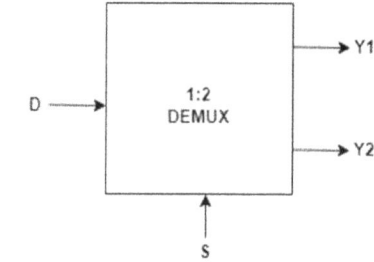

Select	Input	Outputs	
S	D	Y1	Y2
0	0	0	0
0	1	0	1
1	0	0	0
1	1	1	0

FIGURE 2.12   1:2 de-multiplexer.

2.4.2 DE-MULTIPLEXER DESIGNS

The de-multiplexer is a combinational circuit which is used to transfer the data from single input line to any one of the output lines depending on the status of the select lines. It is also called a data distributor and can be utilized to convert serial data into parallel data. The truth table and block diagram for 1:2 de-multiplexer is shown in Figure 2.12.

A number of researchers have studied and proposed various multiplexers using QCA technology, but the research work done on de-multiplexer using QCA is limited. From the literature, some eminent and recent research work done on de-multiplexers using QCA is detailed below.

Shah et al. (2011) proposed a modular 1:2 de-multiplexer with 56 QCA cells in area of 0.08 μm^2 [48]. The designed circuit has latency of 1. With the use of this design, 1:4 and 1:8 de-multiplexers have also been designed which utilized 219 and 613 QCA cells, respectively. **Iqbal et al.** (2013) developed a 1:2 de-multiplexer with an improvement in complexity, latency, and area [42]. Two majority gates and two inverters were used to design the circuit. The designed de-multiplexer uses 27 QCA cells and area around 0.023 μm^2 was utilized. The approach provided a quick procedure for development of higher order de-multiplexers with the use of this proposed 1:2 de-multiplexer. **Safoev and Jeon** (2016) developed 1:2 de-multiplexer as well as 1:4 de-multiplexer circuits [43]. The proposed 1:2 de-multiplexer used two clocks, one inverter and two majority voters. The designed circuit used 21 QCA cells and consumed an area of around 0.02 μm^2. On the other hand, 1:4 de-multiplexer used 140 QCA cells and 0.10 μm^2 area was required for the design. **Das et al.** (2017) suggested a new 1:2 de-multiplexer circuit designing the switching network [44]. The proposed de-multiplexer was designed with three clocks, one inverter and three majority gates.

TABLE 2.5
Summary of literature survey for de-multiplexer

1:2 De-multiplexer Design	Area (In μm²)	Complexity (No. of QCA Cells)	Latency (Clock Cycle)	Circuit Cost = Area * Cells * Latency
[49], 2011	0.08	56	1	4.48
[43], 2013	0.023	27	0.5	0.31
[44], 2016	0.017	21	0.5	0.18
[45], 2017	0.026	32	0.75	6.24
[47], 2017	0.02	21	0.5	0.21
[46], 2018	0.023	21	0.5	0.24
[48], 2021	0.015	19	0.25	0.071

32 QCA cells, 0.75 latency and utilizes the area 0.026 μm². This designed circuit was used as an internal block in the communication network. **Khan et al.** (2017) projected an optimal 1:2 de-multiplexer circuit with the use of two clocks and five inverters, but no majority gate was required in this design [46]. The circuit utilized 21 QCA cells and 0.5 clock latency. The total area required was 0.02 μm². **Ahmad F.** (2018) designed a 1:2 de-multiplexer circuit using 21 QCA cells and utilizing an area of 0.03 μm² [45]. The latency of this circuit was 0.50. Majority voter approach has been used by the author for designing 1:2 de-multiplexer circuit and the same design and approach were used for designing of the 1:4 de-multiplexer circuit, which included 187 QCA cells and utilized 0.18 μm² area. **Sharma** (2021) proposed 1:2 de-multiplexer that uses 19 QCA cells and 0.25 clock latency [47]. The designed circuit requires an area of around 0.015 μm². The circuit has been designed using single layer without any crossover. The comparison of the above studied 1:2 de-multiplexer is summarized in Table 2.5.

2.5   DESIGNING OF FLIP-FLOP

The circuit whose output depends on the present inputs and past output is called a sequential circuit. Thus, the circuit is designed with the use of a combinational circuit along with a memory unit. The basic sequential circuit is flip-flop. Flip-flop is a circuit which is used to store one bit. It holds a state until it is forced to change the state. A basic flip-flop is designed with the use of four-NAND or four-NOR gates. Basically, there are four types of flip-flops, i.e. RS flip-flop, JK flip-flop, D flip-flop and T flip-flop.

2.5.1   D FLIP-FLOP

The most important flip-flop is D flip-flop. D flip-flop also called as Data flip or Delay flip-flop has only one data input 'D' and one clock pulse input. The D flip-flop has two outputs Q and Q'. The output of D flip-flop follows the input with a delay of one clock pulse. Various authors have presented different architectures for the design of D flip-flop using either level triggering or edge triggering clock. The flip-flop can also be operated with edge triggered clock signal which can be further defined as positive edge triggered, negative edge triggered or dual edge triggered flip-flop.

Vetteth, A. during 2003 proposed a level triggered D flip-flop with the involvement of three 3-input majority gates and one inverter gate [65]. The architecture has been designed with the help of 66 QCA cells within an area of 0.08 μm². The drawback of the circuit is unavailability of set/ reset capability and presence of one crossover. **Hashemi, S. and Navi, K.** in 2012 proposed a level triggered D flip-flop with the use of 2:1 multiplexer [66]. The structure has been designed with 48 QCA cells consuming an area of 0.05 μm² and clock latency of 1. Three 3-input majority gates have been employed for designing of the structure. The designed circuit lacks set/ reset capability. **Sasamal, T. N. et al.** in 2018 presented the design of D flip-flop with the help of 2:1 multiplexer which has been proposed by them in the same paper [67]. One 3-input rotated majority gate, one 5-input majority voter and one inverter has been used for the proposed 2:1 multiplexer and further design of level triggered D flip-flop. The proposed D flip-flop consumes an area of around 0.035 μm² with 37 number of QCA cells and the circuit has a clock latency of 0.75. **Trailokya Nath Sasamal et al.** during 2018 devised a 2:1 multiplexer with the help of three 3-input rotated majority gates and one inverter [68]. This suggested 2:1 multiplexer has been utilized by the authors to develop the level triggered D flip-flop. The designed D flip-flop consist of 23 QCA cells covering an area of 0.02 μm² and clock latency of the circuit is 0.5. The circuit has the drawback of non availability of set/ reset facility. **Roshan, MG. et al.** in 2018 proposed both D latch and D flip-flop [70]. The latch was proposed by the authors without the use of any majority gate. The suggested level triggered D flip-flop has been designed with the help of one 3-input majority gate and one inverter. Total of 30 QCA cells consuming an area of 0.02 μm² have been involved in the design of D flip-flop. The design is special as it included the set/ reset facility. **Salma Yaqoob et al.** in 2021 presented a level triggered flip-flop involving three 3-input majority voter and one inverter gate [74]. The complexity of the circuit is 21 QCA cells covering an area of 0.016 μm². The architecture lacks the capability of set/ reset facility.

The summary of the above explained designs for level triggered D flip-flop is presented in Table 2.6.

Hashemi, S. and Navi, K in 2012 proposed a positive edge triggered D flip-flop with the use of 2:1 multiplexer [66]. The structure has been designed with 84 QCA cells consuming an area of 0.09 μm² and clock latency of 2.75. Four 3-input majority gates have been employed for designing of the structure. The circuit has been designed without set/ reset capability. **Sasamal, T. N. et al.** in 2018 presented the design of D flip-flop with the help of 2:1 multiplexer which has been proposed by them in the same paper [67]. One 3-input rotated majority gate, one 5-input majority voter and one inverter has been used for the proposed 2:1 multiplexer and further design of positive edge triggered D flip-flop. The proposed D flip-flop consumes an area of around 0.06 μm² with 59 number of QCA cells and the circuit has a clock latency of 2. **Trailokya Nath Sasamal et al.** during 2018 devised a 2:1 multiplexer with the help of three 3-input rotated majority gates and one inverter [68]. This suggested 2:1 multiplexer has been utilized by the authors to develop the positive edge triggered D

TABLE 2.6
Summary of literature survey for level triggered D flip-flop

| D Flip-Flop Structure | No of Majority Gate | | No. of Inverters Used | No. of Crossover | Set/Reset Ability | Area (In μm²) | Complexity (No. of QCA Cells) | Latency (Clock Cycle) | Circuit Cost = Area * Cells * Latency |
	3-Input MV	3-Input RMG							
[66], 2003	3	0	1	1	NO	0.08	66	1.5	7.92
[67], 2012	3	0	1	0	No	0.05	48	1	2.4
[68], 2018	One 5-Input MV	1	1	0	No	0.035	37	0.75	0.97
[69], 2018	0	3	1	0	No	0.02	23	0.5	0.23
[71], 2018	0	0	0	0	No	0.02	19	0.75	0.29
[71], 2018	1	0	1	0	Yes	0.02	30	1	0.6
[74], 2020						0.012	16	0.5	0.096
[75], 2021	3	0	1	0	No	0.016	21	1	0.34

flip-flop. The designed D flip-flop consist of 47 QCA cells covering an area of 0.04 μm^2 and clock latency of the circuit is 1.75. The circuit lacks the set/ reset facility. **Roshan, MG. et al.** in 2018 proposed both D latch and D flip-flop [70]. The latch was proposed by the authors without the use of any majority gate. The suggested positive edge triggered D flip-flop has been designed with the help of one 3-input majority gate and one inverter. Total of 53 QCA cells consuming an area of 0.04 μm^2 have been involved in the design of D flip-flop. The design is special as it included the set/ reset facility. **Zoka, S. et al.** in 2018 suggested two different designs with the help of one 5-input majority gat [69]. The better design proposed by the authors have involved 95 QCA cells consuming an area of 0.11 μm^2. One crossover has been used in the structure. The architecture has the clock latency of 1 and has set/ reset facility. **Binaei, R. and Gholami** presented a positive edge triggered flip-flop with the employment of six 3-input majority gates and three inverters [71]. The architecture consists of 73 QCA cells that consume an area of 0.1 μm^2. Set/ reset capability is available in the designed circuit. But, the clock latency of the architecture is 2.5, which makes it slightly slow as compared to other designs. **Binaei, R. et al.** in 2019 proposed a positive edge triggered D flip-flop that is based on the four 3-input majority gates and two inverters [72]. A total of 56 QCA cells within an area of 0.06 μm^2 have been utilized in the proposed architecture. The drawback of the circuit is un-availability of set/ reset capability. However, the authors have also proposed the modified structure that includes the set/ reset capability, but the number of used QCA cells involved and area consumed are larger, i.e. 74 QCA cells and 0.09 μm^2, respectively.

The summary of the above explained designs for positive edge triggered D flip-flop is presented in Table 2.7.

Hashemi, S. and Navi, K. in 2012 proposed a negative edge triggered D flip-flop with the use of 2:1 multiplexer [66]. The structure has been designed with 84 QCA cells consuming an area of 0.09 μm^2 and clock latency of 2.75. Four 3-input majority gates have been employed for designing of the structure. The designed circuit does not have the set/ reset capability. **Sasamal, T. N. et al.** in 2018 presented the design of D flip-flop with the help of 2:1 multiplexer which has been proposed by them in the same paper [67]. One 3-input rotated majority gate, one 5-input majority voter and one inverter has been used for the proposed 2:1 multiplexer and further design of negative edge triggered D flip-flop. The proposed D flip-flop consumes an area of around 0.06 μm^2 with 59 number of QCA cells and the circuit has a clock latency of 2. **Trailokya Nath Sasamal et al.** during 2018 devised a 2:1 multiplexer with the help of three 3-input rotated majority gates and one inverter [68]. This suggested 2:1 multiplexer has been utilized by the authors to develop the negative edge triggered D flip-flop. The designed D flip-flop consists of 47 QCA cells covering an area of 0.04 μm^2 and clock latency of the circuit is1.75. The circuit lacks the set/ reset facility. **Roshan, MG. et al.** in 2018 proposed both D latch and D flip-flop [70]. The latch was proposed by the authors without the use of any majority gate. The suggested negative edge triggered D flip-flop has been designed with the help of one 3-input majority gate and one inverter. Total of 53 QCA cells consuming an area of 0.04 μm^2 have been involved in the design of D flip-flop. The circuit has the clock latency of 2.25 and the design is special as it included the set/ reset facility. **Binaei, R. and Gholami** presented a negative edge triggered flip-flop with the employment of six 3-input majority gates

TABLE 2.7
Summary of literature survey for positive edge triggered D flip-flop

D Flip-Flop Structure	No of Majority Gate		No. of Inverters Used	No. of Crossover	Set/Reset Ability	Area (In μm²)	Complexity (No. of QCA Cells)	Latency (Clock Cycle)	Circuit Cost = Area * Cells * Latency
	3-Input MV	3-Input RMG							
[67], 2012	4	0	2	0	No	0.09	84	2.75	20.79
[68], 2018	One 5-Input MV	1	-	0	No	0.06	59	2	7.08
[69], 2018	0	4	2	0	No	0.04	47	1.75	3.29
[70], 2018	1 (5-input MV gate)	0	4	1	Yes	0.11	95	1	10.45
[71], 2018	1	0	1	0	Yes	0.04	53	2.25	4.77
[72], 2019	6	0	3	0	Yes	0.1	73	2.5	18.25
[73], 2019	4	0	2	0	No	0.06	56	2.5	8.4

and three inverters [71]. The architecture consists of 73 QCA cells that consume an area of $0.1 \ \mu m^2$. Set/ reset capability is available in the designed circuit. But, the clock latency of the architecture is 2.5, which makes it slightly slow as compared to other designs. **Binaei, R. et al.** in 2019 proposed a negative edge triggered D flip-flop that is based on the four 3-input majority gates and two inverters [72]. A total of 56 QCA cells within an area of $0.06 \ \mu m^2$ have been utilized in the proposed architecture. The drawback of the circuit is un-availability of set/ reset capability. However, the authors have also proposed the modified structure that includes the set/ reset capability, but the number of used QCA cells involved and area consumed are larger, i.e. 74 QCA cells and $0.09 \ \mu m^2$, respectively.

The summary of the above explained designs for negative edge triggered D flip-flop is presented in Table 2.8.

Hashemi, S. and Navi, K. in 2012 proposed a dual edge triggered D flip-flop with the use of 2:1 multiplexer [66]. The structure has been designed with 120 QCA cells consuming an area of $0.14 \ \mu m^2$ and clock latency of 3.25. Six 3-input majority gates and three inverters have been employed for designing of the structure. The set/ reset capability is not available in the designed circuit. **Sasamal, T. N. et al.** in 2018 presented the design of D flip-flop with the help of 2:1 multiplexer which has been proposed by them in the same paper [67]. One 3-input rotated majority gate, one 5-input majority voter and one inverter has been used for the proposed 2:1 multiplexer and further design of dual edge triggered D flip-flop. The proposed D flip-flop consumes an area of around $0.11 \ \mu m^2$ with 91 number of QCA cells and the circuit has a clock latency of 2.25. **Trailokya Nath Sasamal et al.** during 2018 devised a 2:1 multiplexer with the help of three 3-input rotated majority gates and one inverter [68]. This suggested 2:1 multiplexer has been utilized by the authors to develop the dual edge triggered D flip-flop. The designed D flip-flop consists of 81 QCA cells covering an area of $0.1 \ \mu m^2$ and clock latency of the circuit is 2.25. The circuit lacks the set/ reset facility. **Roshan, MG. et al.** in 2018 proposed both D latch and D flip-flop. The latch was proposed by the authors without the use of any majority gate [70]. The suggested dual edge triggered D flip-flop has been designed without the use of majority gate, but one inverter has been involved. Total of 65 QCA cells consuming an area of $0.05 \ \mu m^2$ have been involved in the design of D flip-flop. The design has the set/ reset facility and clock latency of 2.25. **Binaei, R. and Gholami** during 2019 presented a dual edge triggered flip-flop with the employment of eight 3-input majority gates and four inverters [71]. The architecture consists of 100 QCA cells that consume an area of $0.12 \ \mu m^2$. Set/ reset capability is available in the designed circuit. But, the clock latency of the architecture is 2.5, which makes it slightly slow as compared to other designs. **Binaei, R. et al.** in 2019 proposed a dual edge triggered D flip-flop that is based on the six 3-input majority gates and three inverters [72]. A total of 83 QCA cells within an area of $0.08 \ \mu m^2$ have been utilized in the proposed architecture. The drawback of the circuit is un-availability of set/ reset capability. However, the authors have also proposed the modified structure that includes the set/ reset capability, but the number of used QCA cells involved and area consumed are larger, i.e. 101 QCA cells and $0.13 \ \mu m^2$, respectively.

The summary of the above explained designs for dual edge triggered D flip-flop is presented in Table 2.9.

TABLE 2.8

Summary of literature survey for negative edge triggered D flip-flop

D Flip-Flop Structure	No of Majority Gate		No. of Inverters Used	No. of Crossover	Set/Reset Ability	Area (In μm^2)	Complexity (No. of QCA Cells)	Latency (Clock Cycle)	Circuit Cost = Area * Cells * Latency
	3-Input MV	3-Input RMG							
[67], 2012	4	0	2	0	No	0.09	84	2.75	20.79
[68], 2018	One 5-Input MV	1	-	0	No	0.06	59	2	7.08
[69], 2018	0	4	2	0	No	0.04	47	1.75	3.29
[71], 2018	1	0	1	0	Yes	0.04	53	2.25	4.77
[72], 2019	6	0	3	0	Yes	0.1	73	2.5	18.25
[73], 2019	4	0	2	0	No	0.06	56	2.5	8.4

TABLE 2.9
Summary of literature survey for dual edge triggered D flip-flop

D Flip-Flop Structure	No of Majority Gate		No. of Inverters Used	No. of Crossover	Set/Reset Ability	Area (In μm²)	Complexity (No. of QCA Cells)	Latency (Clock Cycle)	Circuit Cost = Area * Cells * Latency
	3-Input MV	3-Input RMG							
[67], 2012	6	0	3	0	No	0.14	120	3.25	54.6
[68], 2018	One 5-Input MV	1	-	0	No	0.11	91	2.25	22.5
[69], 2018	0	6	3	0	No	0.1	81	2.25	18.2
[71], 2018	0	0	1	0	Yes	0.05	65	2.25	7.3
[72] 2019	8	0	4	0	Yes	0.12	100	2.5	30
[73], 2019	6	0	3	0	No	0.08	83	2.5	16.6

2.5.2 T Flip-Flop

In case of T flip-flop, output Q gets complemented, i.e. changes from logic state 1 to logic state 0 or vice versa, when both T-input and CLK signal are at logic state 1. If the CLK signal is at logic 0, the output holds its previous value. Therefore, in T flip-flop, the output equation is as follows:

$$Q(t+1) = T*CLK*Qt' + (T*CLK)'*Qt$$

where $Q(t+1)$ represents the next state output and Qt represents present state, T represents the current value of input and CLK is clock or enable signal. A number of T flip-flops have been presented in the past by various authors. Some of them have been listed in Table 2.10.

Torabi, M. in 2011 devised a T flip-flop with the involvement of three 3-input majority gates and two inverters [85]. Total area consumed by the 66 QCA cells in the circuit has been found to be 0.06 μm². The clock latency of the circuit is 1.25. **Angizi, S. et al.** during 2014 proposed a level sensitive T flip-flop utilizing 55 QCA cells covering an area of 0.06 μm² [86]. Four 3-input majority voters have been employed for the designing of the architecture. The circuit is slow in terms of producing output as the clock latency of the circuit is 1.5. **Angizi, S. et al.** in 2015 presented a coplanar level triggered T flip-flop for designing counters [87]. The suggested T flip-flop used four 3-input majority gates and two inverters. The architecture was efficient to previous designs in terms of complexity (46 QCA cells), compactness (0.06 μm² area) and speed (clock latency of 1.0). **Majeed, A.H. et al.** in 2019 devised a level sensitive T flip-flop using XOR gate with NAND gate [88]. The suggested T flip-flop has been designed with the use of only 21 QCA cells which consumes an area of 0.018 μm². The designed T flip-flop has been used by the authors to construct synchronous counters.

2.5.3 JK Flip-Flop

In case of JK flip-flop, when both input are at logic 1 and the CLK signal is also at logic state 1, then output Q will get complemented. If the CLK signal is at logic 0, the output holds its previous value.

$$Q(t+1) = CLK*(J*Qt' + K'*Qt) + CLK'*Qt$$

where $Q(t+1)$ represents the next state output and Qt represents present state, J and K represents the current value of input and CLK is clock or enable signal. Various structures of JK flip-flop have been proposed by the authors and researchers in the past. All the authors have attempted to design the circuit with lowest complexity and higher speed. The details of some of them are given in Table 2.11.

Xiaokou Yang et al. in 2010 proposed a falling edge triggered JK flip-flop with the help of 120 QCA cells [89]. The circuit has been found to have the clock latency of 2.0. With the use of this proposed flip-flop, the authors have also designed a synchronous counter. **Yongqiang Zhang et al.** during 2018 presented a dual edge

TABLE 2.10
Summary of literature survey for coplanar T flip-flop

| T Flip-Flop | No of Majority Gate | | No. of Inverters Used | No. of Crossover | Area (In μm^2) | Complexity (No. of QCA Cells) | Latency (Clock Cycle) | Circuit Cost = Area * Cells * Latency |
	3-Input MV	5-Input MV						
[86], 2011	3	0	2	0	0.06	66	1.25	4.95
[87], 2014	4	0	2	0	0.06	55	1.5	4.95
[88], 2015	4	0	2	0	0.06	46	1	2.76
[89], 2019	1	0	0	0	0.0186	21	0.5	0.2

TABLE 2.11
Summary of literature survey for coplanar JK flip-flop

| JK Flip-Flop | No of Majority Gate | | No. of Inverters Used | No. of Crossover | Area (In μm²) | Complexity (No. of QCA Cells) | Latency (Clock Cycle) | Circuit Cost = Area * Cells * Latency |
	3-Input MV	5-Input MV						
[90], 2010	6	0	3	0	0.22	120	2	52.8
[91], 2018	9	0	4	1	0.16	132	1.25	26.4
[92], 2016	4	0	2	0	0.09	54	1.5	7.29

triggered JK flip-flop that employed 132 QCA cells, but the circuit was faster and compact compared to previous structures [90]. The area consumed by the circuit was 0.16 μm^2 and clock latency of the circuit was 1.25. The structure consists of nine 3-input majority gates, four inverters and one crossover. **Shraddha Pandey et al.** during 2016 proposed a level triggered JK flip-flop utilizing four majority gates and two inverters [91]. The circuit has been found to be very efficient in terms of complexity, i.e. it utilizes only 54 QCA cells. The area consumed by the circuit is 0.09 μm^2.

2.6   DESIGNING OF MEMORY CELL

Memory is the basic circuit which is required for saving information. Random Access Memory (RAM) is an essential component of any computer system. RAM is designed with the help of memory cells which are arranged in 2-D array fashion. The data can be accessed from RAM in random manner, due to which the RAM becomes faster than other memory units. There are basically two types of RAM (Random Access Memory) QCA layout. These are: loop-based and line-based. The operational manner determines how the bit of information is stored either using a loop that contains all the four clocking zones or using a QCA line to save the previous value of the output. In literature, various QCA layouts of RAM cell have been reported made up with D flip-flop to hold a bit of information. Some of the designs have set/reset and read/write select capability.

K. Walus during 2003 designed a loop-based RAM cell that utilizes 158 QCA cells consuming an area of around 0.16 μm^2 [75]. However, the proposed memory cell was slow as the clock latency of the circuit was 2, but, this design motivated the researchers to design more efficient RAM cells. **Mostafa et al.** during 2011 proposed two different designs for RAM cell [76]. The second structure proposed by the authors has been found to be more efficient in terms of complexity (63 QCA cells in first structure as compared to 100 QCA cells in first structure) and area consumption as compared to his first proposed structure. The speed of the second structure is also double to first structure. Both the circuits lack the set/ reset capability.

Sara Hashemi et al. in 2012 proposed the RAM cell having set/ reset ability with the use of 2:1 multiplexer proposed by them [77]. The design consists of 109 QCA cells that consume an area of 0.13 μm^2. The clock latency of the circuit is 1.75. **Sasamal T.N. et al.** in 2019 presented an efficient RAM cell with the use of two 2:1 multiplexers [84]. 5-input majority voter and 3-input rotated majority gates have been utilized to construct these multiplexers and finally the RAM cell has been designed utilizing 75 QCA cells consuming an area of 0.098 μm^2. The clock latency of the designed RAM has been found to be 1.5. The suggested memory cell is coplanar with no crossover and has the set/ reset ability that make this design very efficient. **Mohammad Heydari et al.** (2019) proposed a coplanar loop-based RAM cell layout with the use of SR flip-flop [81]. The design consists of 87 QCA cells covering an area of 0.12μm^2. In 2018, **Sasamal T. N et al.** [80] suggested a RAM cell layout with the use of six rotated majority gates. The architecture contains 49 QCA cells and consumes an area of around 0.06 μm^2. The clock latency of the circuit is 1.25. **Angizi, S et al**. in 2015 constructed a RAM cell composed of three 3-input majority voter and one 5-input majority voter [78]. The design is built with set/ reset facility and consists

TABLE 2.12
Summary of literature survey for RAM cell

QCA RAM Cell	No of Majority Gate		No. of Inverters Used	No. of Crossover	Set/Reset Ability	Area (In μm²)	Complexity (No. of QCA Cells)	Latency (Clock Cycle)	Circuit Cost = Area * Cells * Latency
	3-Input MV	3-Input RMG							
[76], 2003	6	0	2	1	No	0.16	158	2	50.56
[77], 2011	6	0	2	1	No	0.11	100	2	22
[77], 2011	5	0	1	0	No	0.07	63	1	4.41
[78], 2012	6	0	2	0	Yes	0.13	109	1.75	24.80
[79], 2015	3 + One 5-MV gate	0	1	0	Yes	0.08	88	1.5	10.56
[80], 2017	3 + One 5-MV gate	0	1	0	Yes	0.06	71	1.25	5.33
[81], 2018	0	6	2	0	Yes	0.06	49	1.25	3.68
[82], 2019	0	6	1	0	Yes	0.12	87	1.25	13.05
[83], 2019				0	Yes	0.13	87	1.5	16.97
[84], 2019	3 + One 5-MV gate	0	1	0	Yes	0.06	67	1.25	5.025
[85], 2019	Two 5-MV gate	2	0	0	Yes	0.098	75	1.5	11.03

of 88 QCA cells occupying an area of 0.08 μm^2. The circuit shows the clock latency of 1.5. **Khosroshahyet al.** in 2017 presented a 5-input majority gate and with the use of it, they further proposed a RAM cell consisting of four 3-input majority voter and one 5-input majority gate [79]. Only 71 QCA cells, consuming an area of 0.06 μm^2 have been used for the designing of the RAM cell. **AzathMubarakali et al.** in 2019 proposed a RAM cell with the use of D-latch proposed by them with the utilization of 2:1 multiplexer [82]. The proposed RAM cell has been designed in coplanar manner with no crossover. It consists of 87 QCA cells and consumes an area of 0.13 μm^2. Output takes 1.5 clock cycles. During 2019, **Ali H. Majeed et al.** suggested a RAM cell which can be designed with the use of three 3-input majority voters, one 5-input majority voter and one inverter [83]. The coplanar design having the set/ reset capability has been designed with the use of 67 cells within an area of 0.06 μm^2, without any crossover.

The summary of past work done on designing efficient RAM cell is given in Table 2.12.

2.7 SUMMARY OF THE CHAPTER

In this paper, QCA technology is presented as an alternative for CMOS technology. Due to the limitations of CMOS technology, QCA technology is becoming the field of interest amongst the researchers. QCA basics such as QCA cell, wire, clocking, logic gates and implementation techniques are studied. QCA along with reversible logics can be the breakthrough in the field of electronics industry. The adders, multiplexers as well as sequential circuits flip-flops and memories which were designed by various researchers with the use of QCA technology have been studied. Utilizing the analysis of this chapter, the researchers of the area will be assisted to develop efficient memories for the future quantum paradigm. However, tremendous research is being done in this field, but more work is yet to be done as there is great potential in this technology.

REFERENCES

1. Singh, R.: "Design and Optimization of Sequential Circuits Using Reversible Logic" September 2020.
2. Tougaw, P. D., and Lent, C. S.: Logical devices implemented using quantum cellular automata. Journal of Applied Physics, 75, 1818–1825 (1994).
3. Landauer, R.: Irreversibility and heat generation in the computing process. IBM Journal of Research and Development, 5, 183–191 (1961).
4. Bennett, C. H.: Logical reversibility of computation. IBM Journal of Research and Development, 17, 525–532 (1973).
5. Moore, G. E.: Lithography and the future of Moore's law, SPIE's 1995 Symposium on Microlithography, International Society for Optics and Photonics (1995).
6. Mack, C.: The multiple lives of Moore's law. IEEE Spectr 52(4), 31–31 (2015).
7. Lent, C. S., Tougaw, P. D., Porod, W., and, Bernstein, G. H.: Quantum cellular automata. Nanotechnology, 4, 49 (1993).
8. Smith, C. G.: Computation without current. Science, 284, 274–274 (1999).

9. Bilal, B., Ahmed, S., and Kakkar, V.: Modular adder designs using optimal reversible and fault tolerant gates in field-coupled QCA nanocomputing. International Journal of Theoretical Physics, 57(5), 1356–1375 (2018).
10. Lent, C. S., et al., Quantum cellular automata. Nanotechnology 4, 49 (1993).
11. Orlov, A. O., Amlani, I. Bernstein, G., Lent, C., and Snider, G.: Realization of a functional cell for quantum-dot cellular automata, Science 277(5328), 928–930 (1997).
12. Kim, K., Wu, K., and Karri, R.: Quantum-dot cellular automata design guideline. IEICE Transactions on Fundamentals of Electronics, Communications and Computer Sciences, E89-A(6), 1607–1614 (2006). https://doi.org/10.1093/ietfec/e89-a.6.1607
13. Zhang, Y., Xie, G., Sun, M., and Lv, H.: An efficient module for full adders in quantum-dot cellular automata. International Journal of Theoretical Physics, 57(10), 3005–3025 (2018).
14. Toth, G. and Lent, Craig S.: Quasiadiabatic switching for metal-island quantum-dot cellular automata. Journal of Applied Physics, 85(5), 51–55 (1999)
15. Lent, Craig S., and Tougaw, P. D.: A device architecture for computing with quantum dots. Proceedings of the IEEE 85(4), (1997), 541–557.
16. Lent, Craig S., and Snider, Gregory L., "The development of quantumdot cellular automata." In: Anderson, N. G., and Bhanja, S. (eds) Field-Coupled Nanocomputing, pp. 3–20. Springer Berlin Heidelberg (2014).
17. Cowburn, R. P., and Welland, M. E., Room temperature magnetic quantum cellular automata. Science 287 (5457), 1466–1468 (2000).
18. Gaur, H. M., et al.: Procedia Computer Science 125, 810–817 (2018).
19. Mohammadi M., and Eshghi, M.: On figures of merit in reversible and quantum logic designs. Quantum Information Processing. 8(4), 297–318 (2009).
20. Sen, B., Goswami, M., Mazumdar, S., and Sikdar, B. K.: Towards modular design of reliable quantum-dot cellular automata logic circuit using multiplexers. Computers & Electrical Engineering, 45, 42–54 (July 2015). https://doi.org/10.1016/j.compeleceng.2015.05.001.
21. Rashidi, H., Rezai, A., and Soltany, S. "High performance multiplexer architecture for quantum-dot cellular automata. Journal of Computational Electronics, 15, 968–981 (September 2016). https://doi.org/10.1007/s10825-016-0832-3.
22. Das, J. C., and De, D.: Optimized multiplexer design and simulation using quantum dot-cellular automata. Indian Journal of Pure & Applied Physics, 54(12), 802–811 (2016).
23. Rashidi, H., and Rezai, A.: Design of novel efficient multiplexer architecture for quantum-dot cellular automata. Journal of Nano and Electronic Physics, 9(1), 01012 (7pp) (2017). https://doi.org/10.21272/jnep.9(1).01012.
24. Asfestani, M. N., and Heikalabad, S. R.: A unique structure for the multiplexer in quantum-dot cellular automata to create a revolution in design of nanostructures. Physica B: Physics of Condensed Matter, 512, 91–99 (May 2017). https://doi.org/10.1016/j.physb.2017.02.028.
25. Khosroshahy, M. B., Moaiyeri, M. H., Angizi, S., Bagherzadeh, N., and Navi, K.: Quantum-dot cellular automata circuits with reduced external fixed inputs. Microprocessors and Microsystems, 50, 154–163 (May 2017). https://doi.org/10.1016/j.micpro.2017.03.009.
26. Ahmad, F.: An optimal design of qca based 2n:1/1:2n multiplexer/De-multiplexer and its efficient digital logic realization. Microprocessors and Microsystems, 56, 64–75 (February 2018). https://doi.org/10.1016/j.micpro.2017.10.010.

27. Ahmadpour, S., and Mosleh, M.: A novel fault-tolerant multiplexer in quantum-dot cellular automata technology. The Journal of Supercomputing, 74(9), 4696–4716 (2018). https://doi.org/10.1007/s11227-018-2464-9.

28. Mosleh, M.: A novel design of multiplexer based on nano-scale quantum-dot cellular automata. Concurrency and Computation: Practice and Experience, 2018 (e5070), 1–16 (2018). https://doi.org/10.1002/cpe.5070.

29. Xingjun, L., Zhiwei, S., Hongping, C., and Haghighi, M. R. J.: A new design of QCA-based nano-scale multiplexer and its usage in communications. International Journal of Communication Systems, 33(4), 1–12 (2019). https://doi.org/10.1002/dac.4254.

30. AlKaldy, E., Majeed, A. H., Zainal, M. S., and Nor, D. B. M.: Optimum multiplexer design in quantum-dot cellular automata. Indonesian Journal of Electrical Engineering and Computer Science, 17(1), 148–155 (2020). doi: 10.11591/ijeecs. v17.i1. pp 148–155.

31. Almatrood, A., George, Aby K., and Singh, H.: Low-Power Multiplexer Structures Targeting Efficient QCA Nanotechnology Circuit Designs. Electronics, 10, 1885 (2021). https://doi.org/10.3390/electronics10161885.

32. Mahesh, M., and Sri. Kumar, T. Vijay: Implementation of Multi-bit Multiplexer Using Majority logic based QCA. Journal of Engineering Science, 13(06), (June 2022)

33. Dahana, S. K., and Hajari, A., Efficient Design of 2:1 MUX _Multiplexer_ using Nanotechnology based on QCA. IJTSRD, 2(6), 1211–1214 (2018).

34. Sen, B., Dutta, M., Saran D., and Sikdar, Biplab K.: "An Efficient Multiplexer in Quantum-dot Cellular Automata." In: Rahaman, H., Chattopadhyay, S., and Chattopadhyay, S. (eds) Progress in VLSI Design and Test, pp. 350–351. Springer (2012).

35. Lee, J. S., and Jeon J. C.: "Design of qca 2-to-1 line multip lexer using NAND Gate." International Journal of Industrial Electronics and Electrical Engineering, 4(6), 51–55 (June 2016). ISSN: 2347-6982.

36. Jeon, J.C.: "Designing nanotechnology qca-multiplexer using majority function-based nand for quantum computing. The Journal of Supercomputing, 77, 1562–1578 (2021).

37. Kianpour, M., and Sabbaghi-Nadooshan, R.: Optimized design of multiplexor by quantum-dot cellular automata. International Journal of Nanoscience and Nanotechnology, 9(1), 15–24 (March 2013).

38. Khan, A., and Arya, R.: Towards the design and analysis of multiplexer/ demultiplexer using quantum dot cellular automata for nano systems. Journal of New Materials for Electrochemical Systems, 25(1), 62–71 (January 2022).

39. Ahmadpour, S. S., Mosleh M., and Heikalabad, S. R.,: Efficient designs of quantum-dot cellular automata multiplexer and RAM with physical proof along with power analysis. The Journal of Supercomputing, 78, 1672–1695 (2022).

40. Sabbaghi-Nadooshan, R., and Kianpour, M., A novel QCA implementation of MUX-based universal shift register. Journal of Computational Electronics, 13, 1–13 (2013).

41. Rashidi, H., and Rezai, A.: Design of novel multiplexer circuits in QCA nanocomputing. Electronics and Energetics, 34(1), 105–114 (March 2021).

42. Iqbal, J. Khanday, F. A., and Shah, N. A.: "Design of quantum-dot cellular automata (QCA) based modular 2n–1–2n MUX-DEMUX. IMPACT 2013, pp. 189–193. IEEE (2013).

43. Safoev, N., and Jeon, J.-C.: Low area complexity de-multiplexer based on multilayer quantum-dot cellular automata. International Journal of Control and Automation, 9(12), 165–178 (2016).

44. Das, J. C., and De, D.: Circuit switching with quantum-dot cellular automata. Nano Communication Networks, 2017; 14, 16–28 (2017).

45. Ahmad, F.: An optimal design of QCA based 2n: 1/1: 2n multiplexer/ de-multiplexer and its efficient digital logic realization. Microprocessors and Microsystems, 56, 64–75 (2018).

46. Khan, A., and Arya, R.: Optimal de-multiplexer unit design and energy estimation using quantum dot cellular automata. Journal of Supercomputing, 77, 1714–1738 (2020).

47. Sharma, V. K.: Optimal design for 1:2 n de-multiplexer using QCA nanotechnology with energy dissipation analysis. International Journal of Numerical Modelling: Electronic Networks, Devices and Fields. doi:10.1002/jnm.2907

48. Shah, N. A., Khanday, F. A., Bangi, Z. A., and Iqbal, J.: Design of quantum-dot cellular automata (QCA) based modular 1 to 2n de-multiplexers. International Journal of Nanotechnology and Applications, 5(1), 47–58 (2011).

49. Gaur, H. M., Sasamal, T. N., Singh, A. K., Mohan, A., and Pradhan, D. K.: Reversible Logic: An Introduction" In A. K. Singh et al. (eds.), Design and Testing of Reversible Logic, Lecture Notes in Electrical Engineering. Springer Nature Singapore Pte Ltd., 577 (2020).

50. Tougaw, P. D., and Lent, C. S.: Logical devices implemented using quantum cellular automata. Journal of Applied Physics, 75(3), 1818–1825 (1994).

51. Kianpour, M., Sabbaghi-Nadooshan, R., and Navi, K.: A novel design of 8-bit adder/ subtractor by quantum-dot cellular automata. Journal of Computer and System Sciences, 80(7), 1404–1414 (2014).

52. Angizi, S., Alkaldy, E., Bagherzadeh, N., and Navi, K.: Novel robust single layer wire crossing approach for exclusive or sum of products logic design with quantum-dot cellular automata. Journal of Low Power Electronics, 10(2), 259–271 (2014).

53. Hashemi, S., and Navi, K.: A novel robust QCA full-adder. Procedia Materials Science, 11, 376–380 (2015).

54. Abedi, D., Jaberipur, G., and Sangsefidi, M.: Coplanar full adder in quantum-dot cellular automata via clockzone-based crossover. IEEE transactions on nanotechnology, 14(3), 497–504 (2015).

55. Sasamal, T. N., Singh, A. K., and Mohan, A.: An efficient design of quantum-dot cellular automata based 5-input majority gate with power analysis. Microprocessors and Microsystems, 59, 103–117 (2018).

56. Sen, B., Rajoria, A., and Sikdar, Biplab K.: Design of efficient full adder in quantum-dot cellular automata hindawi publishing corporation, The Scientific World Journal, 2013, Article ID 250802 (2013).

57. Girija, S., Bellary, P., Monisha, L., Rahul, Sai P., and Rakesh, B.: Full adders using quantum dot cellular automata (QCA). International Journal of Latest Technology in Engineering, Management & Applied Science, VII(III), 135 ISSN 2278-2540 (March 2018). www.ijltemas.in

58. Sadeghi, M., Navi, K., and Dolatshahi, M.: "A New Quantum-Dot Cellular Automata Full-Adder" 5th International Conference on Computer Science and Network Technology (ICCSNT) (2016).

59. Zoka, S., and Gholam, M.: A novel efficient full adder–subtractor in QCA nanotechnology. International Nano Letters 9, 51–54 (2019).

60. Gassoumi, I., Touil, L., and Mtibaa, A.: An efficient design of QCA full-adder-subtractor with low power dissipation. Hindawi, Journal of Electrical and Computer Engineering. 2021, Article ID 8856399, 9 pages (2021).

61. Altarawneh, Z., and Al-Tarawneh, M.: Improved QCA-Based Full Adder/ Subtractor Structures. International Review of Electrical Engineering (I.R.E.E.). 16(4). ISSN 1827–6660 (July–August 2021).

62. Laajimi, R.: "Nano architecture of Quantum-Dot Cellular Automata (QCA) Using Small Area for Digital Circuits." In: Mingbo Niu (ed.) Advanced Electronic Circuits–Principles, Architectures and Applications on Emerging Technologies, InTech, pp. 67–84 (2018).

63. Wang, L., and Yan, J.: An efficient full adder circuit design in Quantum-dot Cellular Automata technology. Advanced Computing in Electron Microscopy, 2020, 1(1), 1–7 (2020).

64. Majeed, A., and Alkaldy, E.: High-performance adder using a new XOR gate in QCA technology. The Journal of Supercomputing, 11564–11579 (2022).

65. Vetteth, A., Walus, K., Dimitrov, V. S., and Jullien, G. A.: "Quantum-Dot Cellular Automata of Flip-Flops." ATIPS Laboratory 2500 University Drive, N.W., Calgary, Alberta, Canada T2N 1N4 (2003).

66. Hashemi, S., and Navi, K.: New robust QCA D flip flop and memory structures. Microelectronics Journal, 43(12), 929–940 (2012).

67. Sasamal, T. N., Singh, A. K., and Ghanekar, U.: Design and implementation of QCA D-flip-flops and RAM cell using majority gates. Journal of Circuits, Systems and Computers, 28(5), 1–19 (2018).

68. Sasamal, T. N., Singh, A. K., and Ghanekar, U.: "Design of QCA-Based D Flip Flop and Memory Cell Using Rotated Majority Gate." Springer Nature Singapore Pte Ltd. (2018)

69. Zoka, S., and Gholami, M.: A novel rising edge triggered resettable D flip-flop using five input majority gate. Microprocessors and Microsystems, 61, 327–335 (2018).

70. Roshan, M. G., and Gholami, M.: Novel D latches and D flip-flops with set and reset ability in QCA nanotechnology using minimum cells and Area. International Journal of Theoretical Physics, 57(Oct (10)):3223–3241 (2018).

71. Binaei, R., and Gholami, M.: Design of novel d flip-flops with set and reset abilities in quantum-dot cellular automata nanotechnology, Computers & Electrical Engineering, 74, 259–272 (2019).

72. Binaei, R., and Gholami, M.: Design of multiplexer-based D flip-flop with set and reset ability in quantum dot cellular automata nanotechnology. International Journal of Theoretical Physics, 58(3), 687–699 (2019).

73. Kumaresan, R. S., Gopalakrishnan, L., and Raj, M.: "Area-Efficient D-Flip Flop and XOR in QCA" 11th ICCCNT 2020 July 1–3, 2020–IIT–Kharagpur.

74. Yaqoob, S., Ahmed, S., Naz, S. F., Bashir S., and Sharma, S.: 3 Design of efficient N-bit shift register using optimized D flip flop in quantum dot cellular automata technology. IET Quantum Communication, 2, 32–41 (2021).

75. Walus, W. Vetteth, A., Jullien, A., and Dimitrov, V. S.: RAM design using quantum-dot cellular automata. Technical Proceedings of the Nanotechnology Conference and Trade Show, 2, 160–163 (2003).

76. Dehkordi, M. A., Shamsabadi, A. S., Ghahfarokhi, B. S., and Vafaei, A.: Novel RAM cell designs based on inherent capabilities of quantum-dot cellular automata. Microelectronics Journal, 42, 701–708 (2011).

77. Hashemi, S., and Navi, K.: New robust QCA D flip flop and memory structures. Microelectronics Journal, 43, 929–940 (2012).

78. Angizi, S., Sarmadi, S., Sayedsalehi, S., and Navi, K.: Design and evaluation of new majority gate-based RAM cell in quantum-dot cellular automata. Microelectronics Journal 46, 43–51 (2015).

79. Khosroshahy, M. B., Moaiyeri, M. H., Navi, K., and Bagherzadeh, N.: An energy and cost efficient majority based RAM cell in quantum-dot cellular automata. Results in Physics, 7, 3543–3551 (2017).

80. Sasamal T. N., Singh A. K., and Ghanekar U.: "Design of QCA-based D Flip flop and memory cell using rotated majority gate." In: Bijaya Ketan Panigrahi, Munesh C. Trivedi, Krishn K. Mishra, Shailesh Tiwari, and Pradeep Kumar Singh (eds) Smart Innovations in Communication and Computational Sciences. Springer, 233–247 (2019).

81. Heydari, M., Xiaohu, Z., Lai, K. K., and Afro, S.: A cost-aware efficient RAM structure based on quantum-dot cellular automata nanotechnology. International Journal of Theoretical Physics, 58 and 3961–3972 (2019)

82. Mubarakali, A., Ramakrishnan, J., Mavaluru, D., Elsir, A., Elsier, O. and Wakil, K.: A new efficient design for random access memory based on quantum dot cellular automata nanotechnology. Nano Communication Networks, 21, 100252 (2019).

83. Majeed, Ali H., AlKaldy E., and Albermany, S.: An energy-efficient RAM cell based on novel majority gate in QCA technology. SN Applied Sciences 1, 1354 (2019).

84. Sasamal, T. N., Singh, A. K., and Ghanekar, U.: Design and implementation of QCA D-Flip-Flops and RAM cell using majority gates. Journal of Circuits, Systems and Computers, 28(5), 1–19 (2019).

85. Torabi, M.: A new Architecture for T Flip Flop Using Quantum-dot Cellular Automata. In Proceedings of the IEEE Asia Symposium on Quality Electronic Design, Kuala Lumpur, Malaysia, pp. 296–300 (19–20 July 2011).

86. Angizi, S., Navi, K., and Sayedsalehi, S.: Efficient quantum dot cellular automata memory architectures based on the new wiring approach. Journal of Computational and Theoretical Nanoscience, 11, 2318–2328 (2014).

87. Angizi, S., Moaiyeri, M., Farrokhi, S., Navi, K., and Bagherzadeh, N.: Designing quantum-dot cellular automata counters with energy consumption analysis. Microprocessors and Microsystems, 39, 512–520 (2015).

88. Majeed, A. H., Alkaldy, E., Zainal, M., and Nor, D. B.: Synchronous counter design using novel level sensitive T-FF in QCA technology. Journal of Low Power Electronics and Applications, 9, 27 (2019).

89. Yang, X., Cai, L., Zhao, X., and Zhang, N.: "Design and simulation of sequential circuits in quantum dot cellular automata: falling edge-triggered flip flop and counter study. Microelectronics Journal 41, 56–63 (2010).

90. Zhang, Y., Xie, G., and Lv, H.: Dual-edge triggered JK flip-flop with comprehensive analysis in quantum-dotcellular automata, Journal of Engineering, 2018(7), 354–359 (2018). doi:10.1049/joe.2018.0138

91. Pandey, S., Singh, S. and Wairya, S.: Designing an efficient approach for JK and T flip-flop with power dissipation analysis using QCA. International Journal of VLSI Design & Communication Systems, 7(3), (June 2016). doi: 10.5121/vlsic.2016.7303 29

92. Gin, A., Williams, S., Meng, H., and Tougaw, P. D.: Hierarchical design of quantum-dot cellular automata devices. Journal of Applied Physics, 85(7), 3713–3722 (1999).

3 An Optimized Approach of Designing Adders and Multiplexer in QCA

Vaibhav Jain[a], Devendra Kumar Sharma[b] and Hari Mohan Gaur[c]
[a]Department of Electronics and Communication Engineering, ABES Institute of Technology, Ghaziabad, India; [a,b]Department of Electronics and Communication Engineering, SRM Institute of Science and Technology, NCR Campus, Ghaziabad, India;
[c]School of Computer Science Engineering and Technology, Bennett University, Greater Noida, India.

3.1 INTRODUCTION

Present CMOS technology starts facing problems such as large power dissipation and feature size reduction due to failure of Denard Scaling [Mack(2015)]. The development at nanoscale requires an alternative technology that helps in the creation of reduced size and low power consumption devices. The QCA (Quantum-Dot Cellular Automata) is the most trending alternative technology [Gaur et al. (2020) Gaur, Sasamal, Singh, Mohan, and Pradhan]. This was invented in 1993 by Lent et al., but practically verified in 1997. The main advantage of QCA lies in the fact that there is no current flow requirement for information propagation. However, in CMOS high power dissipation occurs due to the current transfer between transistors. Reversible circuits is one of the other alternative solutions to reduce power dissipation [Gaur and Singh(2016), Gaur et al. (2015) Gaur, Singh, and Ghanekar]. Unlike CMOS which requires external power supply, QCA uses clock signal which provides the real power to the circuit for its operation [Tougaw and Lent(1994), Tougaw and Lent(1996)].

The fundamental gates in QCA technology are 3-input majority gate, wire and inverter. Several digital architectures have been developed which are combinations of fundamental gates. Presently, these three primitives with clock signal will decide the design complexity. Presently, cell interaction approach is widely used in designing that leads to optimization at larger scale. Arithmetic and logic unit (ALU) is used to perform arithmetic and logical operations in any microprocessor. Therefore, designing of small size adder and subtractor architecture becomes essential inside any processor. [Gaur et al. (2019) Gaur, Singh, and Ghanekar]. This chapter represents

DOI: 10.1201/9781003361633-3

several combinational circuit designs in QCA technology such as half adder, full adder, half subtractor, full subtractor and multiplexer. The proposed half adder takes 25 cells in an area of 0.03 μm^2 and full adder uses 31 cells in 0.02 μm^2 area. The half subtractor design builds with 22 cells and occupies 0.03 μm^2 area whereas utilizing 31 cells in an area of 0.02 μm^2 full subtractor circuits have been presented. A 2 × 1 multiplexer design is also implemented that consists of 22 cells and produces output after 03 clock zones.

In literature, several authors have reported half and full adder structures using different combinations of 3-input majority gate and inverters [Ahmad et al. (2014) Ahmad, Ahmad, and Khan; Lakshmi and Athisha (2011); Jagarlamudi et al. (2011) Jagarlamudi, Saha, and Jagarlamudi; Santra and Roy (2014), Ajitha et al. (2015) Ajitha, Ramanaiah, and Sumalatha; Poorhosseini and Hejazi (2018); Tougaw and Lent(1994); Vetteth et al. (2002) Vetteth, Walus, Dimitrov, and Jullien; Wang et al. (2003) Wang, Walus, and Jullien; Kim et al. (2006) Kim, Wu, and Karri; Bishnoi et al. (2012) Bishnoi, Giridhar, Ghosh, and Nagaraju; Kianpour et al. (2014) Kianpour, Sabbaghi-Nadooshan, and Navi; Angizi et al. (2014) Angizi, Alkaldy, Bagherzadeh, and Navi; Hashemi and Navi (2015); Abedi et al. (2015) Abedi, Jaberipur, and Sangsefidi; Sasamal et al. (2018) Sasamal, Singh, and Mohan]. To achieve minimization, adder designs are also reported that utilized 5-input majority gate [Bishnoi et al. (2012) Bishnoi, Giridhar, Ghosh, and Nagaraju; Angizi et al. (2014) Angizi, Alkaldy, Bagherzadeh, and Navi; Hashemi and Navi (2015); Sasamal et al. (2018) Sasamal, Singh, and Mohan]. The subtractor circuit designs were also reported using 3-input, 5-input majority gate, and inverters [Lakshmi et al. (2010) Lakshmi, Athisha, Karthikeyan, and Ganesh; Dallaki and Mehran (2015); Reshi and Banday (2016); Ahmad et al. (2017) Ahmad, Quadri, Tantary, Wani, Ahmad, and Bahar; Labrado and Thapliyal (2016); Hayati and Rezaei (2015); Jaiswal and Sasamal (2017)]. During past decade, the reported QCA half adder designs utilizes a maximum of 77 cells whereas 34 is the minimum. The latency is found to be varying from a value of 2 to 0.75 in single or multilayer. However, the area coverage by the designs starts with 0.09 μm^2 and ends with 0.05 μm^2. The full adder designs have been reported with a maximum cell count of 292 in an area of 0.62 μm^2 which generates output after 14 clock zones. It can be seen that optimization continuity leads to the development of low size full adder circuit having 46 cells in 0.04 μm^2 area. The design utilizes 04 clock zones to produce the required output which results in reduced cost-function value to 10. The development of half subtractor circuits in QCA takes 62 cells and 08 clock zones initially. However, the optimized design utilizes only 25 cells in an area of 0.03 μm^2 and generates output after 02 clock zones. Similar to full adder, the full subtractor designs have been reported with 178 cells in an area of 0.21 μm^2 which generates output after 08 clock zones. The design is presented with high complexity as it utilizes seven 3-input majority gates with four inverters and four crossovers in single layer. Further optimized designs have been presented with 37 cells that occupies 0.04 μm^2 area. It takes 03 clock zones to produce the output with cost-function value reduced to 1.5. Addition to combinational, sequential circuit designs were also developed and reported in the literature. Multiplexer design used at primary level in the designing of several sequential circuits such as flip-flops, registers and counters [Kim et al. (2006) Kim, Wu, and Karri; Amiri et al. (2008) Amiri, Mahdavi, and Mirzakuchaki; Mardiris and Karafyllidis (2010); Roohi et al. (2011) Roohi, Khademolhosseini, Sayedsalehi, and Navi; Askari and)Taghizadeh (2011; Mukhopadhyay and Dutta (2012); Kianpour and Sabbaghi-Nadooshan (2013); Das and De (2016); Singh et al. (2016) Singh, Pandey, and Wairya]. The development of multiplexers has been found to start with 56 cells to a minimum of 18. The lower speed design takes 04 clock zones while higher speed architecture produces output after 02 clock zones only.

The chapter organization is as follows: Section 2 gives information essential to understand QCA technology. The proposed adder, subtractor and multiplexer design based on cell interaction approach is discussed in Section 3. Also it covers simulation results and comparison with prior work along power dissipation calculation. Section 4 demonstrates circuit reliability computation for each proposed design and the conclusions have been presented in Section 5.

3.2 QCA TERMINOLOGY

This section elaborates the terms essential to understand the QCA technology. It starts with the QCA cell that uses the Coulomb force interaction phenomenon to transfer information from one end to another. Wire, inverter, clocking schemes and the parameters used to calculate the performance of QCA designs are also elaborated in brief.

(1) **QCA Cell:** It consists of four quantum dots which are placed at four respective corners in the form of square as illustrated in Figure 3.1. Also populated with two electrons arranged diagonally due to coulombic repulsion force between them. The electrons can move from one dot to its adjacent dot depending upon the level of potential barrier between dots. Therefore, a cell can be polarized in one of the two possible states as +1 (Logic "1") and −1 (Logic "0") respectively.

(2) **QCA Wire:** This is an arrangement consisting of more than one QCA cell placed in the form of a line. It is used to transfer digital logic state from one cell to another cell as illustrated in Figure 3.2. For example, if the polarization of first cell is at logic '1' then the same will be propagated to the last or output cell.

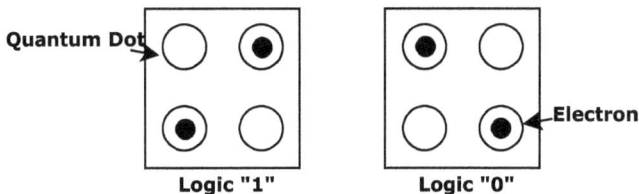

FIGURE 3.1 QCA cell.

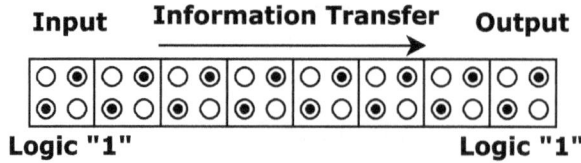

FIGURE 3.2   QCA wire.

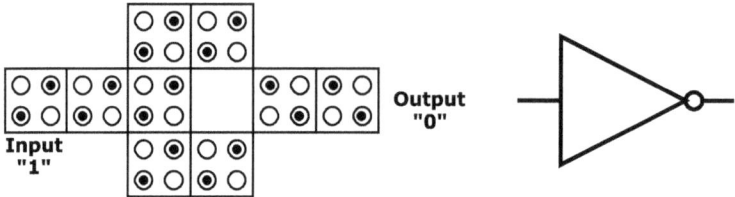

FIGURE 3.3   QCA inverter.

(3) **QCA Inverter:** To perform complement of the input signal, two types of inverter architecture can be found in QCA technology as shown in Figure 3.3. Here, if the input cell is at logic '1' then the corresponding output must be at logic '0' and vice-versa.

(4) **QCA 3-Input Majority Gate:** In QCA technology, basic AND/OR logic operation can be performed by arranging the cells as illustrated in Figure 3.4. The layout is called 3-input majority gate (MV3) and consists of 05 cells. Three are used as inputs, one works as device and remaining one is used to generate output. The Boolean function of MV3 is governed by Eq. 1.

$$MV\,3_{QCA} = AB + BC + AC \tag{1}$$

where A, B and C represents input cells. Out of these three, any one cell is either fixed at logic "1" to perform OR operation or fixed at logic "0" to make AND gate. The symbol and respective truth table that shows all the possible input and output combinations has been illustrated in Figure 3.4 and Table 3.1 respectively.

(5) **QCA Clocking:** It is found that no current flow exist in QCA designs. Therefore a clock signal is essential to control the flow of information and synchronization. Eventually, the clock supplies real power to simulate the circuit [Tougaw and Lent(1994), Tougaw and Lent(1996)]. As per literature, adiabatic switching is a widely utilized clocking scheme in circuit designing in which a clock cycle divides into four clock phases/zones called Switch, Hold, Release, and Relax respectively as illustrated in Figure 3.5. Each clock zone is differ from its adjacent zone by phase difference of $\pi/2$ [Lent and Tougaw (1997)].

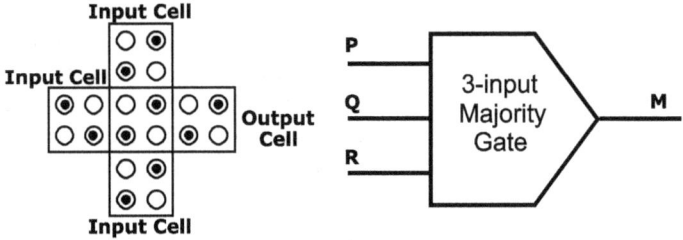

FIGURE 3.4   QCA 3-input majority gate.

TABLE 3.1
Truth table of 3-input majority gate

Input			Output
A	B	C	OUT
0	0	0	0
0	0	1	0
0	1	0	0
0	1	1	1
1	0	0	0
1	0	1	1
1	1	0	1
1	1	1	1

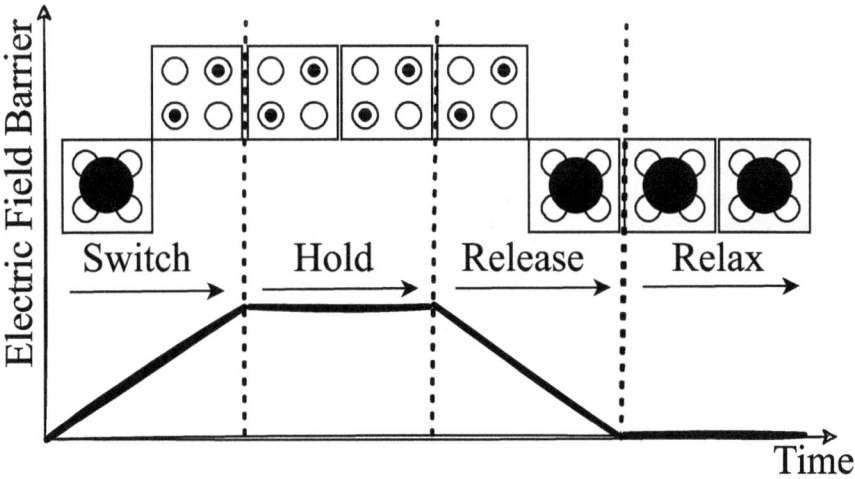

FIGURE 3.5   Energy barrier vs time diagram in QCA clocking.

(6) **Performance Parameters:** In QCA technology, the proposed circuit layout can be compared with the prior work on the basis of some measures called as cell count, area, latency and cost function. Cell count refers to the number of cells utilized, whereas the total space covered by the cell is called area. Latency is calculated with the formula (0.25 * total clock phases utilized to generate the required output). Cost function defines complexity of design and depends on number of majority gates, inverters and crossover used. [Liu et al. (2014) Liu, Lu, ONeill, and Swartzlander]. For a layout design, it is calculated using the formula defined in Eq. 2.

$$Cost_{QCA} = (M^k + I + C^l) * T^p \qquad (2)$$

where M indicates the number of majority gates, I are the inverters, C represents crossover and T shows latency of the circuit. Also, k, l and p are coefficients. In general, the value of k and l is 2 and the value of p is 1.

(7) **Power Dissipation:** The aim behind requirement of change in technology is not only fulfilled by reducing circuit area and delay. Power dissipation measurement of design architecture is also one of the important limitations in CMOS technology. Therefore, the two kinds of power called total power and average power per cycle have been calculated in QCA technology. The simulation tool utilized for this measurement is popularly known as QCA Designer-E which works in coherence vector energy simulation environment [Torres et al. (2018) Torres, Wille, Niemann, and Drechsler]. This tool initialized with 500 000 samples but only 3000 will be taken and recorded for making graph. The simulation results provides the information regarding total and average power per cycle, number of iterations required to converge the initial steady state polarization, and simulation time.

3.3 FEW COMBINATIONAL LOGIC CIRCUITS IN QCA

As per literature, more than a hundred combinational logic circuits have been already designed and presented in QCA technology [Sasamal et al. (2020) Sasamal, Gaur, Singh, and Mohan; Sasamal et al. (2016) Sasamal, Singh, and Mohan; Angizi et al. (2014) Angizi, Alkaldy, Bagherzadeh, and Navi; Hashemi and Navi (2015); Abedi et al. (2015) Abedi, Jaberipur, and Sangsefidi; Sasamal et al. (2018) Sasamal, Singh, and Mohan]. These involve the design of half adder, full adder, half subtractor, full subtractor, multiplexer, decoder, comparator, etc. These circuits can be designed using any of three approaches named as cell interaction, majority gate and reversible gate. This section presents the design of adder, subtractor and multiplexer using cell interaction approach in which the layout of previously reported Exclusive-OR gate is utilized [Jain et al. (2022) Jain, Sharma, and Gaur].

3.3.1 HALF ADDER

A half adder (HA) circuit is utilized for the addition of 2-bits in digital electronics. In general, there are two outputs in addition operation called SUM and CARRY.

Considering A and B as inputs, the respective outputs are governed using one 2-input EX-OR gate and a 3-input majority gate performing AND operation as shown in Figure 3.6. The Boolean equation by which outputs are calculated is defined in Eq. 3 and Eq. 4 respectively.

$$SUM_{HA} = A \oplus B \tag{3}$$

$$CARRY_{HA} = AB \tag{4}$$

The truth table that includes all the possible input combinations and their respective outputs has been listed in Table 3.2. Its proposed equivalent QCA layout can be found in Figure 3.7 which consists of 25 cells in an area of 0.03 μm^2. The circuit utilizes only 02 clock zones to produce output, hence latency of the circuit is 0.5, also verified with the simulation results demonstrated in Figure 3.8. The designed HA layout is also simulated on a tool called QCA Designer-E for the measurement of power dissipation.

It takes 06 iterations to converge the initial steady state polarization and 11 seconds as total simulation time with 02 inputs and 02 output. The amount of total and average energy per cycle during the simulation has been tabulated in Table 3.3. It can be found in literature that several authors presented half adder designs with variations of complexities and design parameters. These are summarized and

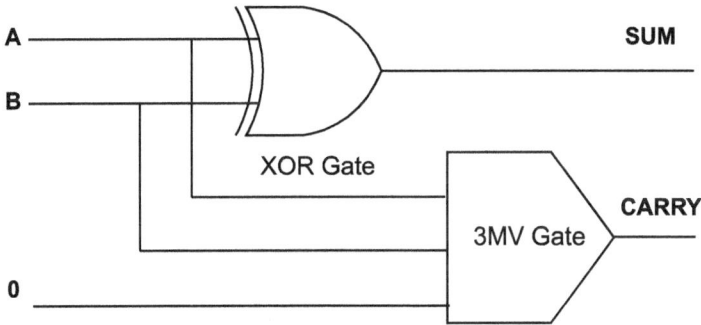

FIGURE 3.6   Logic diagram of half adder.

TABLE 3.2
I/O behavior of half adder

Input		Output	
A	B	SUM	CARRY
0	0	0	0
0	1	1	0
1	0	1	0
1	1	0	1

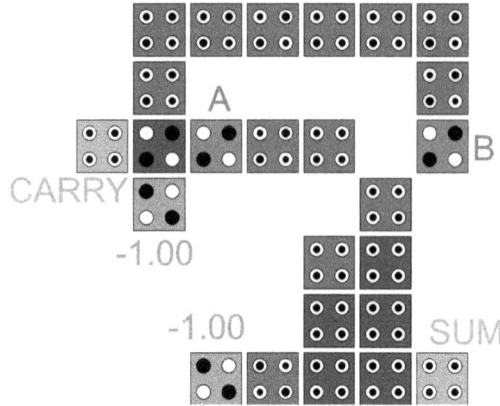

FIGURE 3.7   QCA layout of half adder.

TABLE 3.3
Half adder: power dissipation

Energy Category	Energy Dissipation Value in eV
Total energy dissipation (Sum Ebath)	1.40e-002
Average energy dissipation per cycle (Avg Ebath)	1.27e-003

compare against performance measures area, cell count, delay, and cost function as tabulated in Table 3.4.

It can be seen from Table 3.4 that the proposed structure proves its supremacy over other structures in area, cell count and delay. However, the cost function of the architecture is similar to HA-6 that results in 1. The design HA-1 and HA-2 utilizes three 03-input majority gates with two inverters. They consists of 62 and 77 cells, occupies an area 0.08 μm^2 and generates output after 08 and 04 clock zones respectively. Utilizing four 03-input majority gates and two inverters, HA-3 and HA-4 have been reported. HA-3 uses 61 cells in an area of 0.07 μm^2 whereas HA-4 utilize 64 cells in 0.09 μm^2 area. The two layouts have different speeds and produce output after 03 and 05 clock zones respectively. Reducing the number of inverter and crossover, HA-5 has been reported with 34 cells, 0.05 μm^2 area, and produces output after 03 clock zones only. The optimized design HA-6 have low cost function of value 1 due to utilization of cell interaction approach. It consists of 44 cell in an area of 0.05 μm^2.

3.3.2   FULL ADDER

A 1-bit full adder (FA) circuit is utilized for the addition of three binary bits. Here, similar to half adder circuit there are two outputs called as SUM and CARRY. Considering *A*, *B* and *C* as inputs, then the outputs are generated using one 3-input

Simulation Results

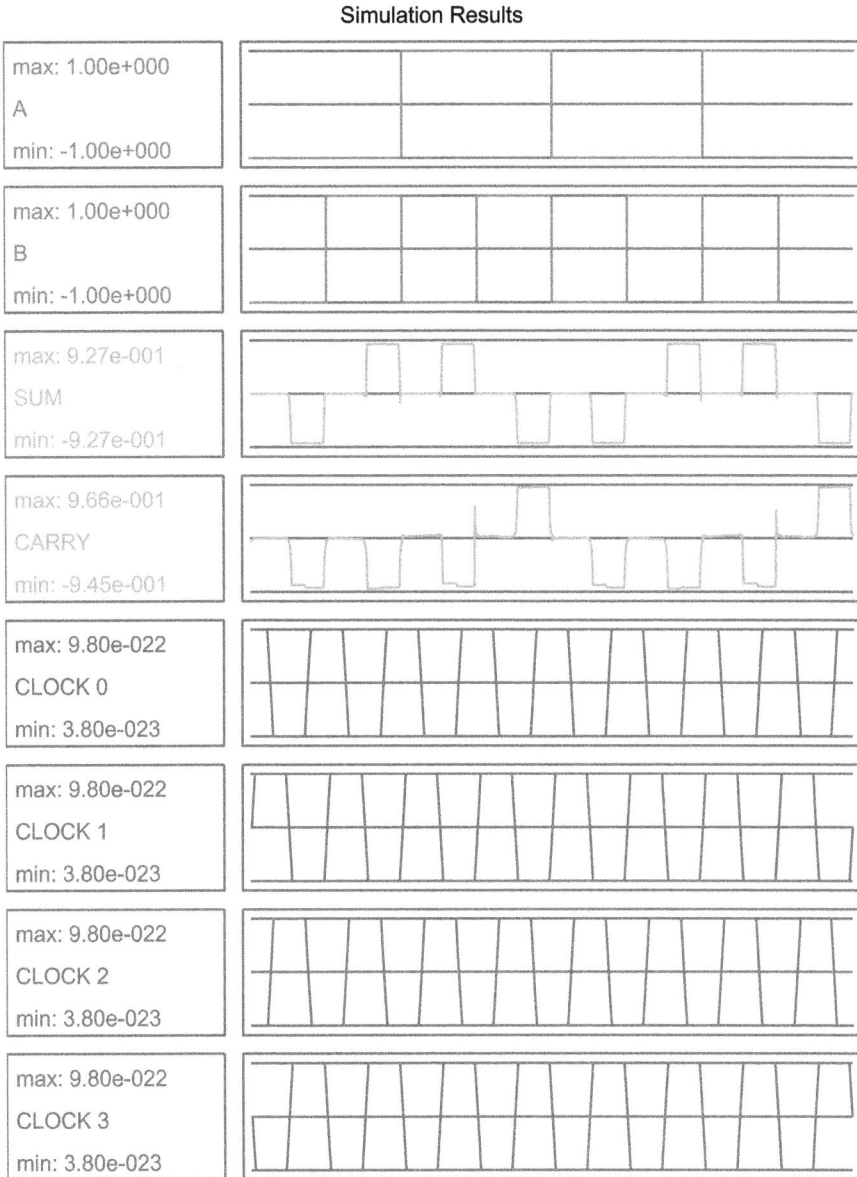

max: 1.00e+000 A min: -1.00e+000	
max: 1.00e+000 B min: -1.00e+000	
max: 9.27e-001 SUM min: -9.27e-001	
max: 9.66e-001 CARRY min: -9.45e-001	
max: 9.80e-022 CLOCK 0 min: 3.80e-023	
max: 9.80e-022 CLOCK 1 min: 3.80e-023	
max: 9.80e-022 CLOCK 2 min: 3.80e-023	
max: 9.80e-022 CLOCK 3 min: 3.80e-023	

FIGURE 3.8   I/O waveform of half adder.

EX-OR gate with a 3-input majority gate as shown in Figure 3.9. The outputs are calculated with the help of Boolean equation defined in Eq. 5 and Eq. 6 respectively.

$$SUM_{FA} = A \oplus B \oplus C \tag{5}$$

$$CARRY_{FA} = AB + BC + AC \tag{6}$$

TABLE 3.4
Comparison QCA half adders

QCA Half Adder Designs	Area Occupied (μm^2)	No. of Cells	Delay (Clock Cycles)	Single/ Multi-layer	Cost Function
HA-1 [Ahmad et al. (2014) Ahmad, Ahmad, and Khan]	0.08	62	2	Single	24
HA-2 [Lakshmi and Athisha (2011)]	0.08	77	1	Multi-layer	12
HA-3 [Jagarlamudi et al. (2011) Jagarlamudi, Saha, and Jagarlamudi]	0.07	61	0.75	Single	13.5
HA-4 [Santra and Roy (2014)]	0.09	64	1.25	Single	22.5
HA-5 [Ajitha et al. (2015) Ajitha, Ramanaiah, and Sumalatha]	0.05	34	0.75	Single	7.5
HA-6 [Poorhosseini and Hejazi (2018)]	0.05	44	1	Single	1
Proposed	0.03	25	0.5	Single	0.5

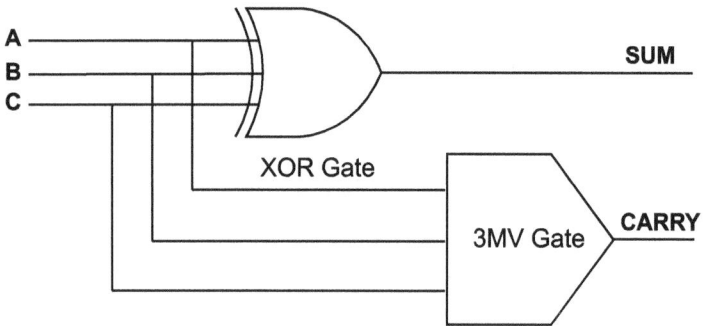

FIGURE 3.9   Logic diagram of full adder.

The truth table showing all possible input and output combinations has been tabulated in Table 3.5. The proposed layout design is in a single layer which consists of 31 cells in an area of 0.02 μm^2 and produces output after 02 clock zones only as represented in Figure 3.10. The simulation results that verify the operation of 1-bit full adder is illustrated in Figure 3.11.

The power dissipation measurement is done using QCA Designer-E which takes 12 iterations to converge the initial steady state polarization. The total simulation time is 17 seconds with 03 inputs and 02 output. The total and average energy per cycle generated in simulation has been listed in Table 3.6. As per literature, several authors

TABLE 3.5
I/O behavior of full adder

Input			Output	
A	B	C	SUM	CARRY
0	0	0	0	0
0	0	1	1	0
0	1	0	1	0
0	1	1	0	1
1	0	0	1	0
1	0	1	0	1
1	1	0	0	1
1	1	1	1	1

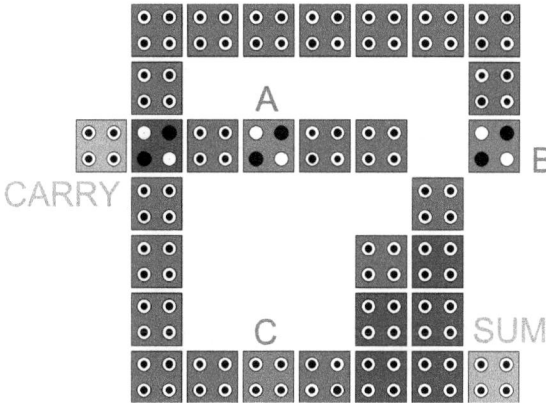

FIGURE 3.10   QCA layout of full adder.

reported full adder designs with variations of complexities and design parameters. The metrics are summarized and compare against previous work as illustrated in Table 3.7.

It can be seen from Table 3.7 that the proposed architecture shows its efficacy over prior structures in all performance metrics; area, cell count, delay, and cost function. At the beginning, utilizing five 03-input majority gates with three inverters FA-1 and FA-2 have been reported. The design uses a large number of crossover which increases the complexity, as a result cost-function becomes very high 109 and 322 respectively. FA-1 consists of 192 cells in 0.20 μm^2 area and uses 04 clock zones to produce output. However, FA-2 builds up with 290 cells in an area of 0.62 μm^2 with 14 clock zones. Further optimized design FA-3, FA-4, FA-6, and FA-9 utilized three 03-input majority gate and two inverters with 6, 3, 2, and 2 crossover. The result is reduction in complexity and performance metrics. FA-3 uses 105 cells, occupied area of 0.17 μm^2 and 05 clock zones to generate the required

Simulation Results

max: 1.00e+000 A min: -1.00e+000	
max: 1.00e+000 B min: -1.00e+000	
max: 1.00e+000 C min: -1.00e+000	
max: 9.27e-001 SUM min: -9.27e-001	
max: 9.51e-001 CARRY min: -9.50e-001	
max: 9.80e-022 CLOCK 0 min: 3.80e-023	
max: 9.80e-022 CLOCK 1 min: 3.80e-023	
max: 9.80e-022 CLOCK 2 min: 3.80e-023	
max: 9.80e-022 CLOCK 3 min: 3.80e-023	

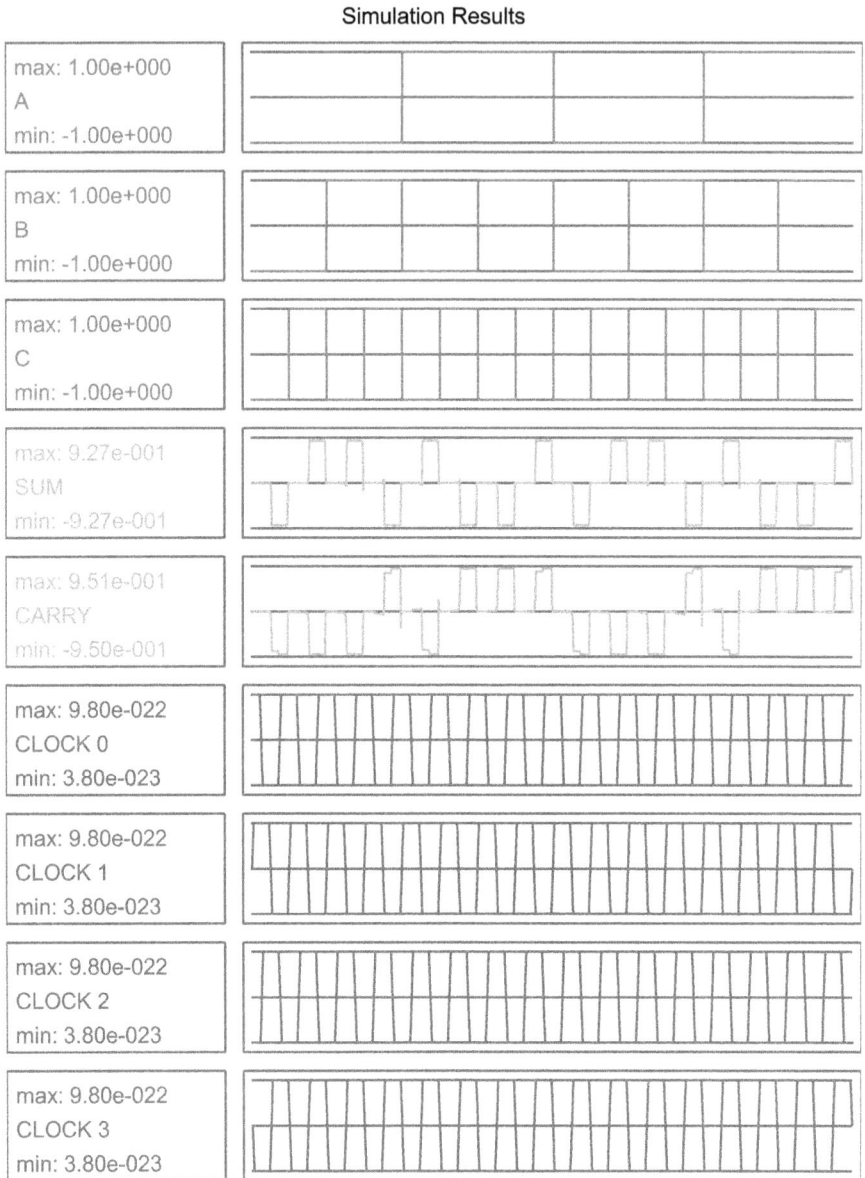

FIGURE 3.11   I/O waveform of full adder.

TABLE 3.6
Full adder: power dissipation

Energy Category	Energy Dissipation Value in eV
Total energy dissipation (Sum Ebath)	1.85e-002
Average energy dissipation per cycle (Avg Ebath)	1.68e-003

TABLE 3.7
Comparison of QCA full adders

QCA Full Adder Designs	Area Occupied (μm2)	No. of Cells	Delay (Clock Cycles)	Single/ Multi-layer	Cost Function
FA-1 [Tougaw and Lent (1994)]	0.2	192	1	Single	109
FA-2 [Vetteth et al. (2002) Vetteth, Walus, Dimitrov, and Jullien]	0.62	292	3.5	Single	322
FA-3 [Wang et al. (2003) Wang, Walus, and Jullien]	0.17	105	1.25	Single	58.75
FA-4 [Kim et al. (2006) Kim, Wu, and Karri]	0.36	220	3	Single	60
FA-5 [Bishnoi et al. (2012) Bishnoi, Giridhar, Ghosh, and Nagaraju]	0.087	95	2	Single	42
FA-6 [Kianpour et al. (2014) Kianpour, Sabbaghi-Nadooshan, and Navi]	0.07	69	1	Single	15
FA-7 [Angizi et al. (2014) Angizi, Alkaldy, Bagherzadeh, and Navi]	0.09	95	1.25	Single	11.25
FA-8 [Hashemi and Navi (2015)]	0.06	71	1.5	Single	15
FA-9 [Abedi et al. (2015) Abedi, Jaberipur, and Sangsefidi]	0.043	59	1	Single	15
FA-10 [Sasamal et al. (2018) Sasamal, Singh, and Mohan]	0.035	46	1	Single	10
Proposed	0.02	31	0.5	Single	0.5

output. Design FA-4 consists of 220 cells in an area of 0.36 μm^2, takes 12 clock zones for the calculation of output. FA-6 have been reported with 69 cells whereas FA-9 represented with 59 cells only. Both designs produces output after 04 clock zones, occupies 0.07 μm^2 and 0.043 μm^2 area respectively. To reduce complexity, FA-5, FA-7, FA-8, and FA-10 have been reported with one 3-input and one 5-input majority gate with either one or two inverters. FA-5 and FA-7 both utilize 95 cells, however they have different speed as FA-5 produces output after 08 clock zones while FA-7 takes only 05 clock zones to produce the required output. As compared to FA-7, cell count has been reduced in FA-8 but it took 06 clock cycle for the generation of output. Therefore, FA-8 design is optimized but slower with respect to FA-7. It is also found that FA-10 utilizes only 46 cells, occupies 0.04 μm^2 area and takes 04 clock zones for output generation.

3.3.3 HALF SUBTRACTOR

A subtractor circuit is equally important to perform arithmetic operations inside a digital processor. A half subtractor (HS) is utilized to generate the difference between 2-bits. Similar to addition, there are again two possible outcomes in subtraction operation called difference(DIFF) and BORROW. Considering A and B as inputs, then required outputs are achieved using one 2-input EX-OR gate, one 3-input majority gate with an inverter as illustrated in Figure 3.12. The Boolean function that defines the required output is given in Eq. 7 and Eq. 8 respectively.

$$DIFF_{HS} = A \oplus B \qquad (7)$$

$$BORROW_{HS} = A'B \qquad (8)$$

The truth table showing all the input combinations and respective output values is listed in Table 3.8. The proposed single layer layout presented in Figure 3.13 utilizes 22 cells, covers an area of 0.03 μm^2 with a delay value of 0.5. The design is based on cell interaction approach, therefore the cost function results in value 1 only. The operation of the circuit is also verified from its simulation results as illustrated in Figure 3.14. The proposed layout of half subtractor is also simulated on QCA

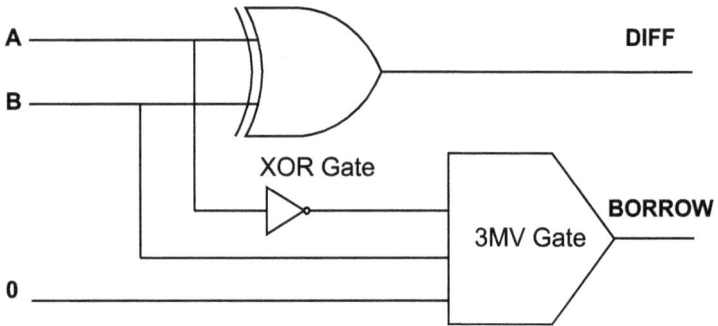

FIGURE 3.12   Logic diagram of half subtractor.

TABLE 3.8
I/O behavior of half subtractor

Input		Output	
A	B	DIFF	BORROW
0	0	0	0
0	1	1	1
1	0	1	0
1	1	0	0

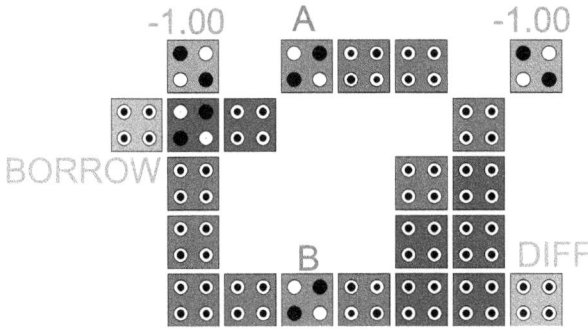

FIGURE 3.13   QCA layout of half subtractor.

Designer-E for power estimation. It starts with 02 inputs and 02 outputs, takes 04 iterations to converge the initial steady state polarization. The total simulation time is found to be 12 seconds. The estimation of total and average energy per cycle has been tabulated in Table 3.9. It can be found in literature that some authors presented half subtractor designs with variations of complexities and design parameters. These are summarized and compare against performance measures area, cell count, delay, and cost function as tabulated in Table 3.10.

It can be observed from Table 3.10 that the proposed architecture is minimized in terms of cell count and cost function. The design runs with same speed and occupies area same as HS-5, also both designs adopted cell interaction design approach. However, HS-5 utilized 25 cells in an area of 0.03 μm^2 with 02 clock zones to produce output. The lower speed design HS-1 uses 62 cells and 08 clock zones to generate the required output. It occupies a total area of 0.08 μm^2 and uses three 03-input majority gate with two inverters and one crossover. With similar number of majority gates, inverters and crossover, HS-2 have been reported with 77 cells in single layer. The design occupies a larger area as compared to HS-1, however it utilizes only 03 clock zones to govern the required output. Utilizing four 3-input majority gate and two inverters, HS-3 is presented. It consists of 55 cells, covers 0.05 μm^2 area and produces output after 03 clock zones. Further optimized design HS-4 utilizes one 5-input majority gate with two inverters and 3-input majority gates. It is built up using 45 cells and produces output after 03 clock zones. The cost function of HS-4 reduces to 8.25 since no usage of crossover as compared to HS-1, HS-2, and HS-3.

3.3.4 FULL SUBTRACTOR

A full subtractor (FS) circuit is utilized to find the difference between three binary bits. Similar to half subtractor, the circuit has two possible outcomes called difference (DIFF) and BORROW. Consider A, B and C as inputs, the corresponding outputs are generated using one 3-input EX-OR gate, one 3-input majority gate with an inverter as represented in Figure 3.15. The outputs Boolean function are generated as per Eq. 9 and Eq. 10 respectively.

Simulation Results

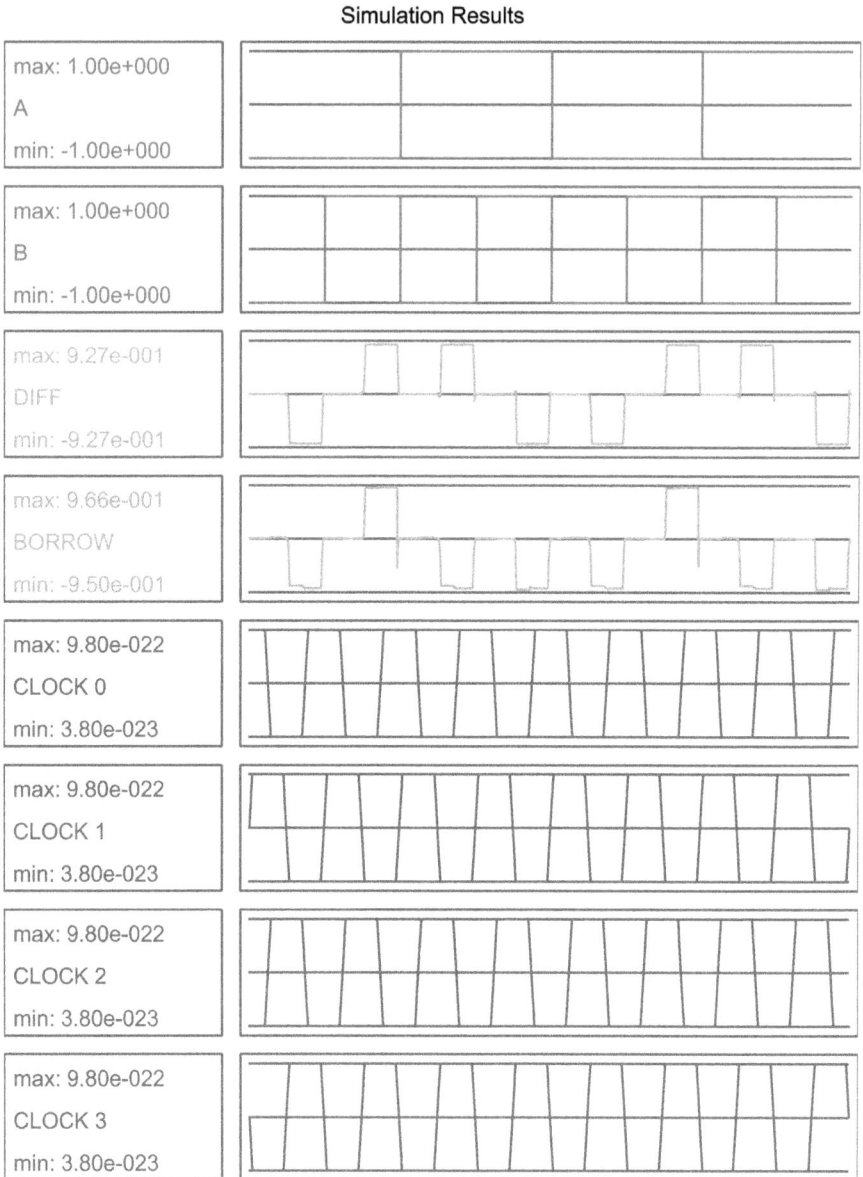

FIGURE 3.14 I/O waveform of half subtractor.

$$DIFF_{FS} = A \oplus B \oplus C \tag{9}$$

$$BORROW_{FS} = A'B + BC + A'C \tag{10}$$

The truth table which governs the output functions with all input combinations has been listed in Table 3.11. The proposed coplanar full subtractor architecture utilizes

TABLE 3.9
Half subtractor: power dissipation

Energy Category	Energy Dissipation Value in eV
Total energy dissipation (Sum Ebath)	1.42e-002
Average energy dissipation per cycle (Avg Ebath)	1.29e-003

TABLE 3.10
Comparison QCA half subtractor

QCA Half Subtractor Designs	Area Occupied (μm^2)	No. of Cells	Delay (Clock Cycles)	Single/ Multi-layer	Cost Function
HS-1 [Ahmad et al. (2014) Ahmad, Ahmad, and Khan]	0.08	62	2	Single	24
HS-2 [Lakshmi et al. (2010) Lakshmi, Athisha, Karthikeyan, and Ganesh]	0.09	77	0.75	Single	9
HS-3 [Dallaki and Mehran (2015)]	0.05	55	0.75	Single	14.25
HS-4 [Reshi and Banday (2016)]	0.04	45	0.75	Single	8.25
HS-5 [Ahmad et al. (2017) Ahmad, Quadri, Tantary, Wani, Ahmad, and Bahar]	0.03	25	0.5	Single	1.5
Proposed	0.03	22	0.5	Single	1

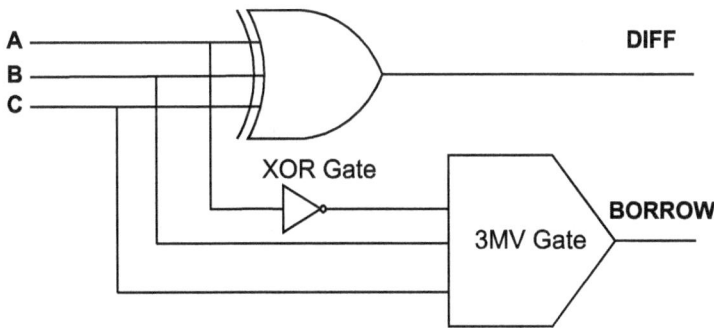

FIGURE 3.15   Logic diagram of full subtractor.

31 cells and occupies 0.02 μm^2 area as shown in Figure 3.16. Its cost function comes out to value 1 and takes 02 clock zones for output generation. The circuit functionality is verified from the simulation results as demonstrated in Figure 3.17. For the measurement of power, the proposed architecture is simulated using QCA designer-E tool.

TABLE 3.11
I/O behavior of full subtractor

Input			Output	
A	B	C	DIFF	BORROW
0	0	0	0	0
0	0	1	1	1
0	1	0	1	1
0	1	1	0	1
1	0	0	1	0
1	0	1	0	0
1	1	0	0	0
1	1	1	1	1

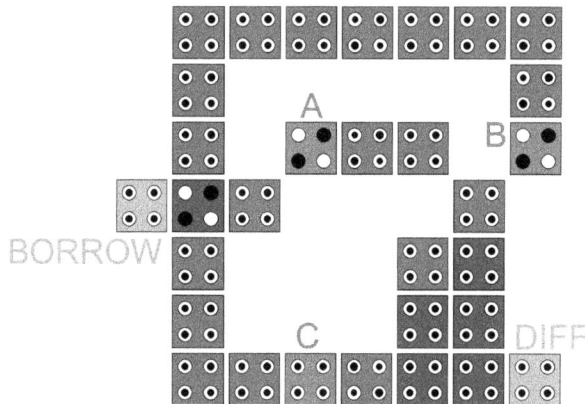

FIGURE 3.16   QCA layout of full subtractor.

It initiates with 03 inputs and 02 outputs, takes total simulation time as 18 seconds. Also, 05 iterations are performed to converge the initial steady state polarization. The total and average energy per cycle during simulation results has been tabulated in Table 3.12. In literature, few designs of full subtractor can be found with variations of complexities and design parameters. These are summarized and compared against performance measures area, cell count, delay, and cost function in single layer or multilayer as tabulated in Table 3.13.

It can be observed from Table 3.13 that the proposed architecture is supreme over prior designs in performance metrics area, cell count, delay, and cost-function. Based on cell interaction method, FS-5 is also reported with 37 cells in an area of 0.04 μm^2 and produces output after 03 clock zones. Utilizing seven 3-input majority gates with four inverters and three crossovers, FS-1 have been reported. It consists of 178 cells in an area of 0.21 μm^2 and takes 08 clock zones to produce the required output. FS-2 is optimized and faster as compared to FS-1 but uses nine 3-input majority gate

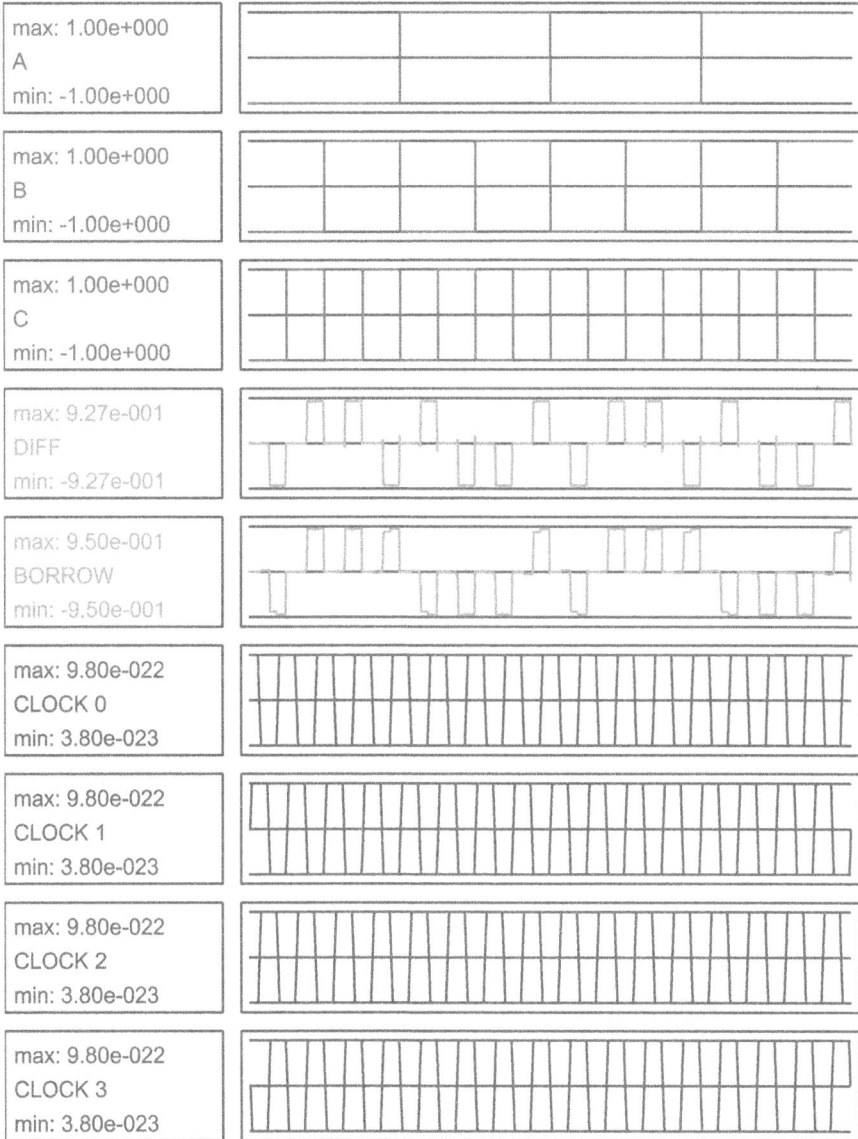

FIGURE 3.17   I/O waveform of full subtractor.

TABLE 3.12
Full subtractor: power dissipation

Energy Category	Energy Dissipation Value in eV
Total energy dissipation (Sum Ebath)	1.62e-002
Average energy dissipation per cycle (Avg Ebath)	1.47e-003

TABLE 3.13
Comparison QCA full subtractor

QCA Full Subtractor Designs	Area Oc cupied (μm^2)	No. of Cells	Delay (Clock Cycles)	Single/ Multi-layer	Cost Function
FS-1 [Lakshmi et al. (2010) Lakshmi, Athisha, Karthikeyan, and Ganesh]	0.205	178	2	Single	124
FS-2 [Dallaki and Mehran (2015)]	0.168	136	1.75	Single	155.75
FS-3 [Reshi and Banday (2016)]	0.146	104	1.75	Single	94.5
FS-4 [Labrado and Thapliyal (2016)]	0.05	63	0.75	Single	12
FS-5 [Ahmad et al.(2017) Ahmad, Quadri, Tantary, Wani, Ahmad, and Bahar]	0.04	37	0.75	Single	1.5
FS-6 [Hayati and Rezaei (2015)]	0.039	52	2	Multi-layer	110
FS-7 [Jaiswal and Sasamal (2017)]	0.05	53	0.75	Single	7.5
Proposed	0.02	31	0.5	Single	1

which increases design complexity and cost-function simultaneously. FS-6 utilizes four 3-input majority gate with three inverters, 52 cells in an area of 0.04 μm^2. The design is optimized in area but lower in speed as it takes 08 clock zones to produce the corresponding outputs. To reduce complexity, 5-input along with 3-input majority gate have been utilized by FS-3, FS-4, and FS-7. FS-3 takes 104 cells in an area of 0.15 μm^2, FS-4 uses 63 cells and occupies 0.05 μm^2 area. However, the optimized FS-7 consists of 53 cells and occupies an area of 0.05 μm^2. Here, FS-3 takes 07 clock zones whereas FS-4 and FS-7 used only 03 clock zones to produce the required results.

3.3.5 MULTIPLEXER

Multiplexer (MUX) is widely used to select any one input as output from 2^n digital inputs. The selection depends upon the state of n lines called selection lines. For example, if n is equal to one then the size of multiplexer becomes 2×1 which shows that there are two input lines, one output and one selection line respectively. Considering A and B as inputs, *SEL* as selection line then the output Y is generated using one 2-input.

EX-OR gate with one 3-input majority gate is shown in Figure 3.18. Here, if input *SEL* = 0 than the value available at input line A will transfer towards output. However, for input line *SEL* = 1 it shows that output Y follows the value at input B. The Boolean expression defining the relation between inputs and outputs is shown in Eq. 11. The proposed layout as shown in Figure 3.19 consists of 22 cells in an area of 0.03 μm^2 and produces output after 03 clock zones in single layer.

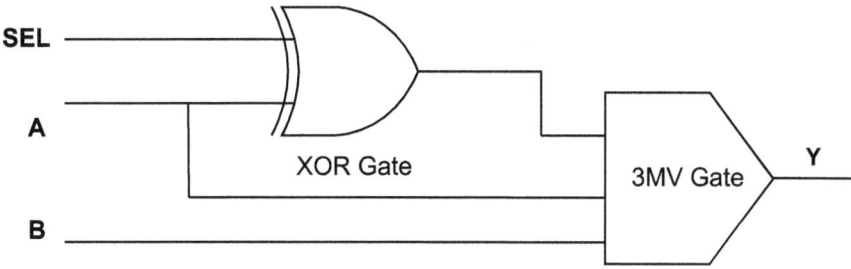

FIGURE 3.18 Logic diagram of 2 × 1 multiplexer.

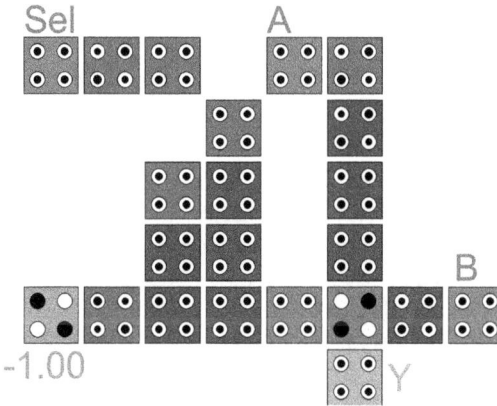

FIGURE 3.19 QCA layout of 2 × 1 multiplexer.

$$Y_{MUX} = (SEL)^t * A + (SEL) * B \tag{11}$$

The truth table listing the possible outcomes with respect to different states of inputs and selection line has been tabulated in Table 3.14. The proposed QCA architecture of 2 × 1 MUX consists of 22 cells in an area of 0.03 μm^2 has been illustrated in Figure 3.19. The latency of layout is found as 0.75 which shows that the output is generated after 03 clock zones. The simulation results that verifies circuit operation is shown in Figure 3.20.

The power estimation is done using QCA designer-E tool which initiates with 03 inputs and 01 output, takes total simulation time as 11 seconds. The iterations required to converge the initial steady state polarization is found to be 09. The total and average energy per cycle findings in simulation have been listed in Table 3.15. Several designs of MUX have been reported in literature with variations of design parameters and complexities. These are tabulated and compared conversely in terms of performance metrics area, cell count, delay, and cost function as tabulated in Table 3.16.

It can be observed from Table 3.16 that the proposed architecture is cost efficient as compared to previous designs. MUX-1, MUX-2, and MUX-3 has been constructed using

TABLE 3.14
I/O behavior of multiplexer

Input			Output
SEL	A	B	Y
0	0	0	0
0	0	1	0
0	1	0	1
0	1	1	1
1	0	0	0
1	0	1	1
1	1	0	0
1	1	1	1

three 3-input majority gate and one inverter. The designs utilized 41, 34 and 56 cells in an area of $0.08\,\mu m^2$, $0.033\,\mu m^2$ and $0.07\,\mu m^2$ respectively. They have same speed and produce output after 04 clock zones. However, the cost-function of MUX-3 results in higher value of 14 due to crossover used in design. MUX-5 reported with three 3-input majority gate with three inverters and takes 04 clock zones to produce the output. It consists of 34 cells in an area of $0.033\,\mu m^2$ with cost function 12. The designs MUX-4, MUX-6, MUX-7, and MUX-8 are a little bit faster as compared to others as they took only 03 clock zones to produce the required output. MUX-4 used 27 cells in $0.03\,\mu m^2$ area, MUX-6 utilized 23 cells in $0.18\,\mu m^2$ area. MUX-7 consists of 22 cells in an area of $0.03\,\mu m^2$ whereas MUX-8 is optimized and built up with 21 cells in $0.02\,\mu m^2$ area. Utilizing three 3-input majority gate with a inverter, high speed MUX-9 and MUX-10 has been reported. The designs takes only 02 clock zones to generate the required output. MUX-9 uses 19 cells, MUX-10 takes only 18 cells and both occupies an area of $0.02\,\mu m^2$.

3.4 PTM AND RELIABILITY FRAMEWORK

Probabilistic Transfer Matrix (PTM) is used to compute the overall reliability of designed circuit at gate level. First, it is calculated for each gate/component including wire segment, FAN-OUT using Kronecker and matrix-matrix multiplication. Thereafter, each component PTM are multiplied together to obtain the overall circuit PTM and reliability [Dysart and Kogge (2007)]. The process involves in the generation of individual component PTM and ETM is as follows:

(1) The ideal transfer matrix (ITM) has been obtained considering both error free and erroneous output in the truth table.
(2) PTM (Probability Transfer Matrix) is generated by replacing the logic "0" value by p (error rate) and logic "1" value by error rate $(1 - p)$.
(3) Erroneous Transfer Matrix (ETM) is obtained from the multiplication of PTM with ITM to remove erroneous (non-desired) terms.

Simulation Results

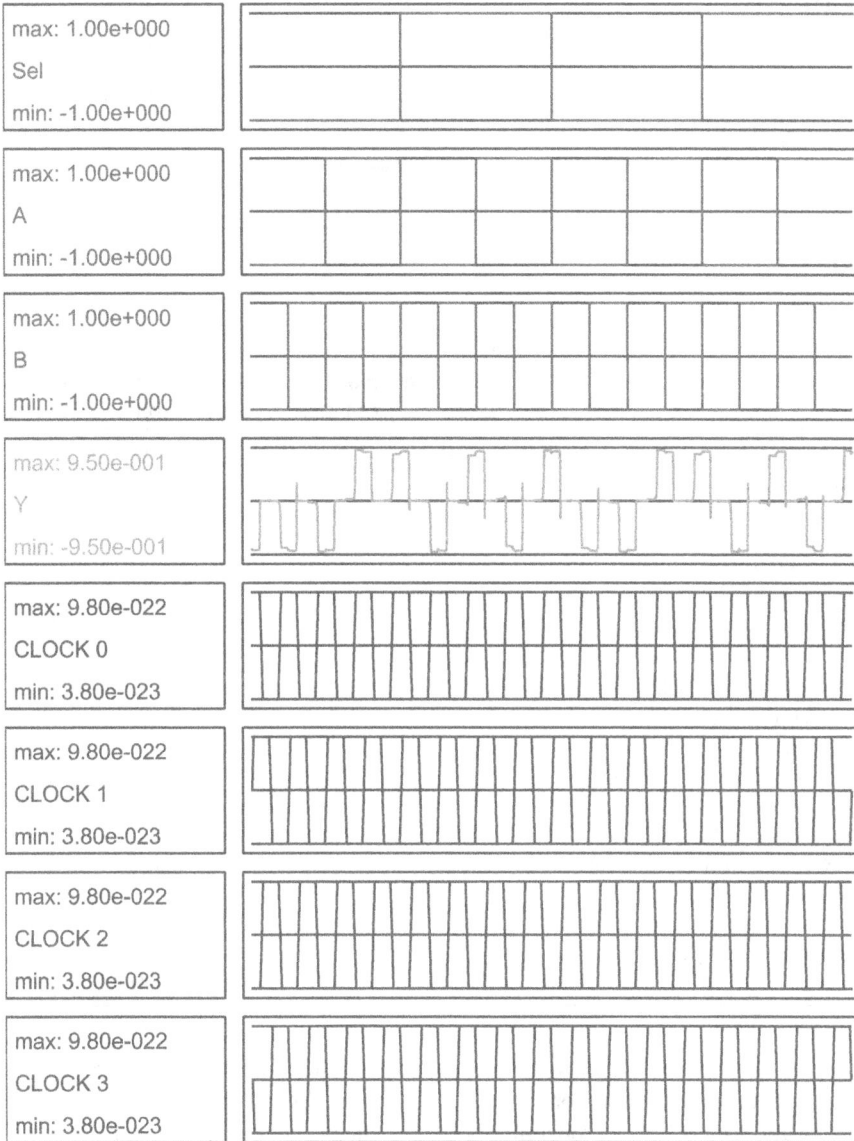

FIGURE 3.20　I/O waveform of 2 × 1 multiplexer.

Using the defined process, calculated PTM of 3-input majority gate, 2-input XOR gate, 3-input XOR gate and wire segment is shown in Table 3.17, Table 3.18, Table 3.19 and Table 3.20 respectively.

The overall reliability of circuit is calculated with components PTM by the following steps:

TABLE 3.15
2 × 1 Multiplexer: Power Dissipation

1. Energy Category	2. Energy Dissipation Value in eV
Total energy dissipation (Sum Ebath)	1.15e-002
Average energy dissipation per cycle (Avg Ebath)	1.04e-003

TABLE 3.16
Comparison QCA 2 × 1 multiplexer

iii. QCA 2 × 1 MUX Designs	Area Occupied (μm^2)	No. of Cells	Delay (Clock Cycles)	Single/ Multi– layer	Cost Function
MUX-1 [Kim et al. (2006) Kim, Wu, and Karri]	0.08	41	1	Single	10
MUX-2 [Amiri et al. (2008) Amiri, Mahdavi, and Mirzakuchaki]	0.033	34	1	Single	10
MUX-3 [Mardiris and Karafyllidis (2010)]	0.07	56	1	Single	14
MUX-4 [Roohi et al. (2011) Roohi, Khademolhosseini, Sayedsalehi, and Navi]	0.03	27	0.75	Single	7.5
MUX-5 [Askari and Taghizadeh (2011)]	0.033	34	1	Single	12
MUX-6 [Mukhopadhyay and Dutta (2012)]	0.18	23	0.75	Single	8.25
MUX-7 [Kianpour and Sabbaghi-Nadooshan (2013)]	0.03	22	0.75	Single	7.5
MUX-8 [Das and De (2016)]	0.015	21	0.75	Single	7.5
MUX-9 [Singh et al. (2016) Singh, Pandey, and Wairya]	0.02	19	0.5	Single	5
MUX-10 [Singh et al. (2016) Singh, Pandey, and Wairya]	0.02	18	0.5	Single	5
Proposed	0.03	22	0.75	Single	0.75

(1) Construct input row vector whose each element defines the probability of input combination. For example, a 2-input XOR gate have 04 input combination, each with a probability of 0.25. Hence its input vector (v) is 4×1 matrix with each element having a value 0.25.

(2) Perform matrix-matrix product of v and component ETM called resultant matrix. For instance, v*ETM for 2-input XOR gate results in [0.5(1 − p) 0.5(1 − p)].

(3) Add the elements value row-wise in resultant matrix to obtain the overall reliability of the circuit. For example, (1 − p) is the final reliability of a 2-input XOR gate.

This section further describes the reliability calculation of proposed half adder, full adder, half subtractor, full subtractor, and multiplexer circuit.

TABLE 3.17
PTM of 3-input majority gate

(a) Truth Table (ITM)

Input			0	1
0	0	0	1	0
0	0	1	1	0
0	1	0	1	0
0	1	1	0	1
1	0	0	1	0
1	0	1	0	1
1	1	0	0	1
1	1	1	0	1

(b) PTM

0	1
(1-p)	p
(1-p)	p
(1-p)	p
p	(1-p)
(1-p)	p
p	(1-p)
p	(1-p)
p	(1-p)

(c) ETM

0	1
(1-p)	0
(1-p)	0
(1-p)	0
0	(1-p)
(1-p)	0
0	(1-p)
0	(1-p)
0	(1-p)

TABLE 3.18
PTM of 2-input XOR gate

(a) Truth Table (ITM)

Input		0	1
0	0	1	0
0	1	0	1
1	0	0	1
1	1	1	0

(b) PTM

0	1
(1-p)	p
p	(1-p)
p	(1-p)
(1-p)	p

(c) ETM

0	1
(1-p)	0
0	(1-p)
0	(1-p)
(1-p)	0

TABLE 3.19
PTM of wire segment

(a) Truth Table (ITM)

Input	0	1
0	1	0
1	0	1

(b) PTM

0	1
(1-p)	p
p	(1-p)

(c) ETM

0	1
(1-p)	0
0	(1-p)

TABLE 3.20
PTM of 3-input XOR gate

(a) Truth Table (ITM)

Input			0	1
0	0	0	1	0
0	0	1	0	1
0	1	0	0	1
0	1	1	1	0
1	0	0	0	1
1	0	1	1	0
1	1	0	1	0
1	1	1	0	1

(b) PTM

0	1
(1-p)	p
p	(1-p)
p	(1-p)
(1-p)	p
p	(1-p)
(1-p)	p
(1-p)	p
p	(1-p)

(c) ETM

0	1
(1-p)	0
0	(1-p)
0	(1-p)
(1-p)	0
0	(1-p)
(1-p)	0
(1-p)	0
0	(1-p)

3.4.1 COMPUTATION FOR HALF ADDER

To compute reliability of proposed half adder, the circuit is divided into sub levels designated as L1, L2, L3, and L4 as illustrated in Figure 3.21. Using Kronecker product and matrix multiplication the PTM of individual level is calculated as per Eq. 12–15.

$$L1 = WireA \oplus WireB \oplus WireC \qquad (12)$$

$$L2 = FANA \oplus FANB \oplus I2 \qquad (13)$$

$$L3 = XOR2 \oplus I8 \qquad (14)$$

$$L4 = MAJ3 \oplus I2 \qquad (15)$$

Here, I2 and I8 are the identity matrix having size 2×2, 8×8 respectively. WireA, WireB, and WireC represents PTM of three inputs A, B, and fixed input "0" respectively. FANA and FANB are FAN-OUT matrix, each of size 2×4 and defined as FAN-OUT = [1 0 0 0; 0 0 0 1]. XOR2 and MAJ3 are the PTM for 2-input XOR gate and 3-input majority gate. The overall PTM of HA is calculated with matrix multiplication of all levels as defined in Eq. 16. Utilizing MATLAB, the computation is done and reliability of the circuit is found for varying error rates as shown in Figure 3.22.

$$PTM_{HA} = L1 * L2 * L3 * L4 \qquad (16)$$

3.4.2 COMPUTATION FOR FULL ADDER

To compute reliability of proposed full adder, the circuit is divided into sub levels designated as L1, L2, L3, and L4 as illustrated in Figure 3.23. Using Kronecker

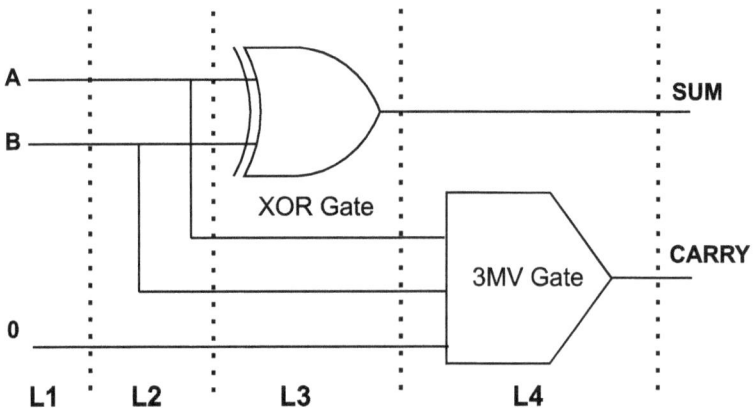

FIGURE 3.21 Half adder divided into 4 levels for PTM calculation.

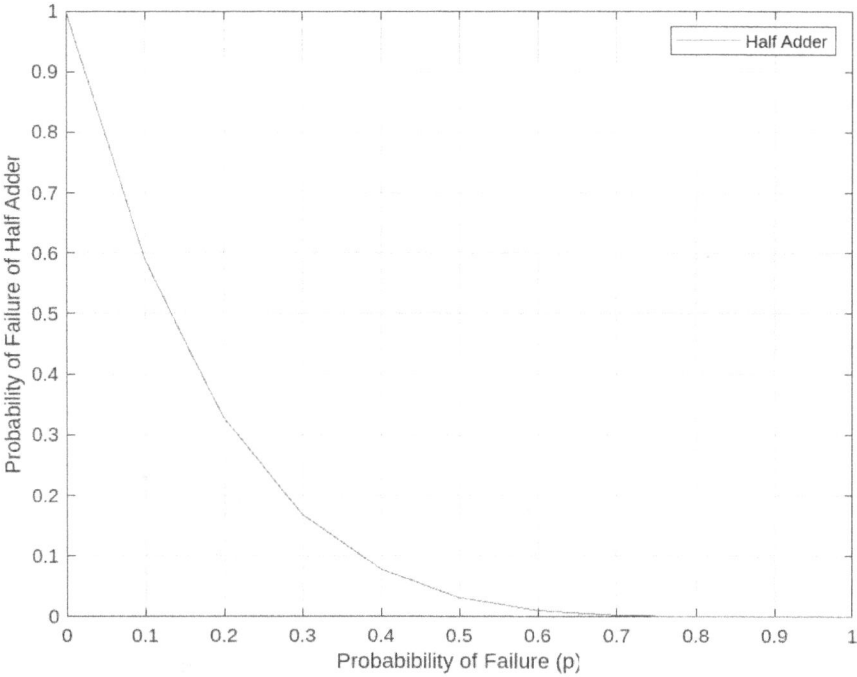

FIGURE 3.22   Reliability of half adder with respect to probability of failure.

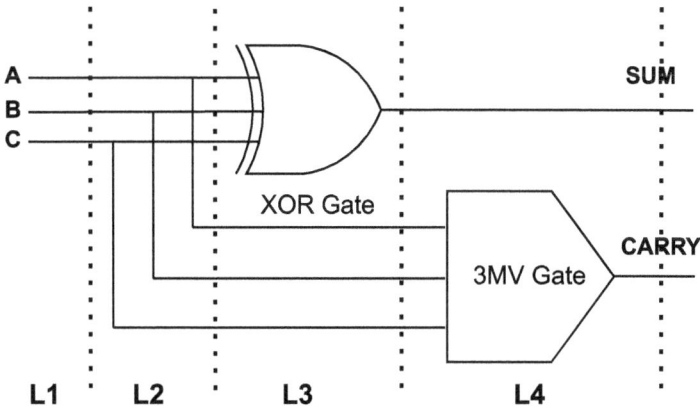

FIGURE 3.23   Full adder divided into 4 levels for PTM calculation.

product and matrix multiplication the PTM of individual level is calculated as per Eq. 17–20.

$$L1 = WireA \oplus WireB \oplus WireC \qquad (17)$$

$$L2 = FANA \oplus FANB \oplus FANC \qquad (18)$$

$$L3 = XOR3 \oplus I8 \tag{19}$$

$$L4 = MAJ \oplus I2 \tag{20}$$

Here, I2 and I8 are the identity matrix having size 2×2, 8×8 respectively. WireA, WireB, and WireC represents PTM of three inputs A, B, and C respectively. FANA, FANB, and FANC are FAN-OUT matrix, each of size 2×4 and defined as FAN-OUT= [1 0 0 0; 0 0 0 1]. XOR3 and MAJ3 are the PTM for 3-input XOR gate and 3-input majority gate. The overall PTM of FA is calculated with matrix multiplication of all levels as defined in Eq. 21. Utilizing MATLAB, the computation is done and reliability of the circuit is found for varying error rates as shown in Figure 3.24.

$$PTM_{FA} = L1 * L2 * L3 * L4 \tag{21}$$

3.4.3 COMPUTATION FOR HALF SUBTRACTOR

To compute reliability of proposed half subtractor, the circuit is divided into sub levels designated as L1, L2, L3, and L4 as illustrated in Figure 3.25. Using Kronecker

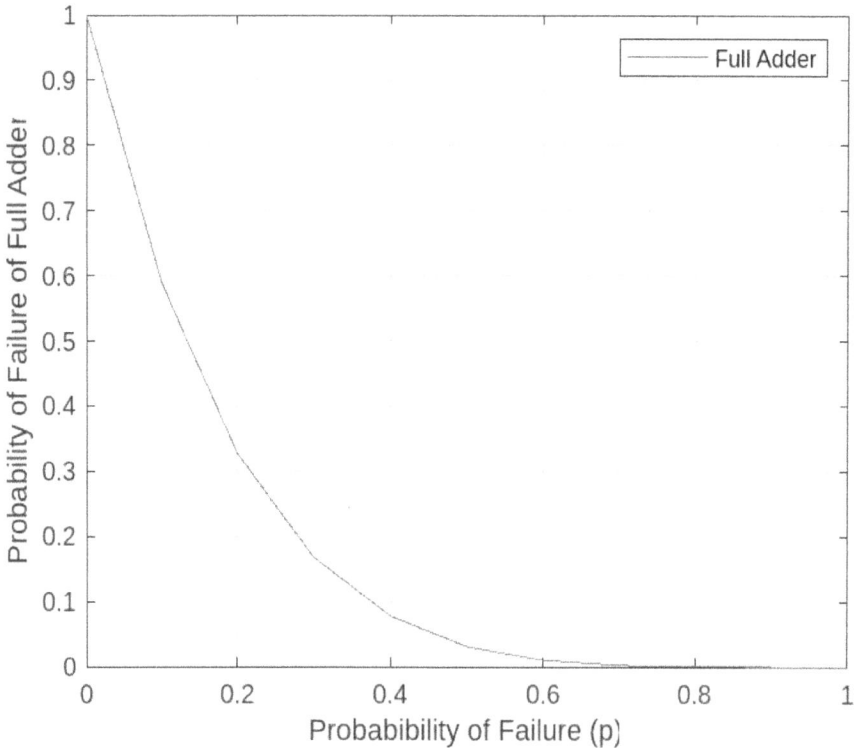

FIGURE 3.24   Reliability of full adder with respect to probability of failure.

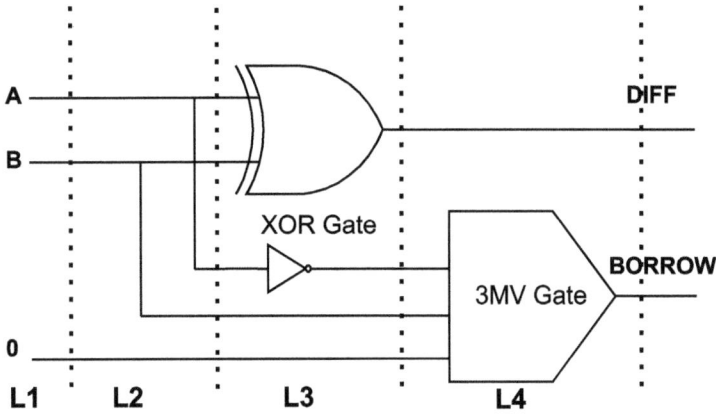

FIGURE 3.25   Half subtractor divided into 4 levels for PTM calculation.

product and matrix multiplication the PTM of individual level is calculated as per Eq. 22–25.

$$L1 = WireA \oplus WireB \oplus WireC \qquad (22)$$

$$L2 = FANA \oplus FANB \oplus I2 \qquad (23)$$

$$L3 = XOR2 \oplus NOT \oplus I4 \qquad (24)$$

$$L4 = MAJ3 \oplus I2 \qquad (25)$$

Here, I2 and I4 are the identity matrix having size 2×2, 4×4 respectively. WireA, WireB and WireC represents PTM of three inputs A, B and fixed input "0" respectively. FANA and FANB are FAN-OUT matrix, each of size 2×4 and defined as FAN-OUT = $[1\ 0\ 0\ 0;\ 0\ 0\ 0\ 1]$. XOR2, NOT and MAJ3 are the PTM for 2-input XOR gate, inverter and 3-input majority gate. The overall PTM of HS is calculated with matrix multiplication of all levels as defined in Eq. 26. Utilizing MATLAB, the computation is done and reliability of the circuit is found for varying error rates as shown in Figure 3.26.

$$PTM_{HS} = L1 * L2 * L3 * L4 \qquad (26)$$

3.4.4   COMPUTATION FOR FULL SUBTRACTOR

To compute reliability of proposed full subtractor, the circuit is divided into sub levels designated as L1, L2, L3, and L4 as illustrated in Figure 3.27. Using Kronecker product and matrix multiplication the PTM of individual level is calculated as per Eq. 27–30.

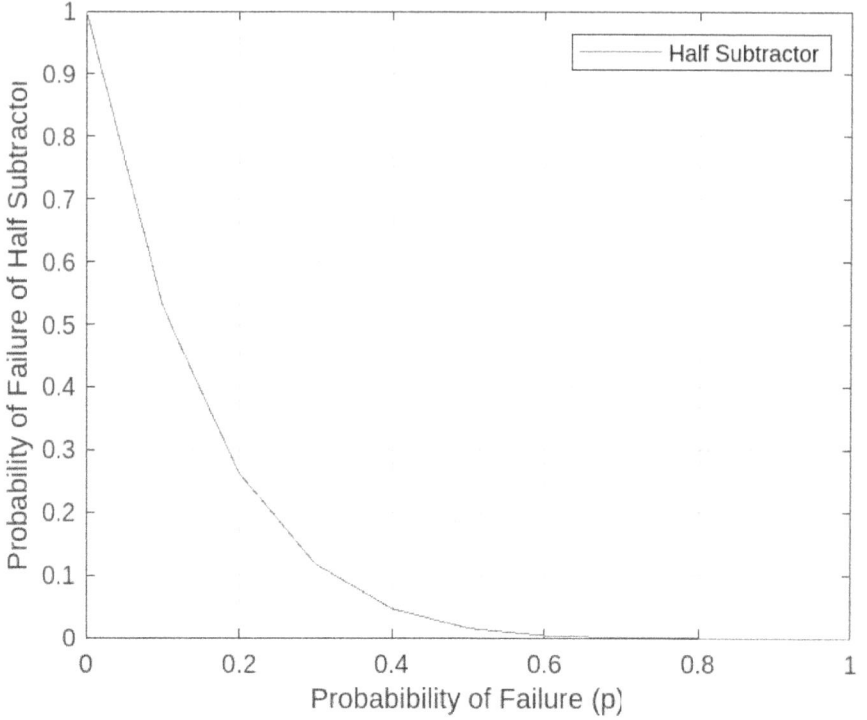

FIGURE 3.26   Reliability of half subtractor with respect to probability of failure.

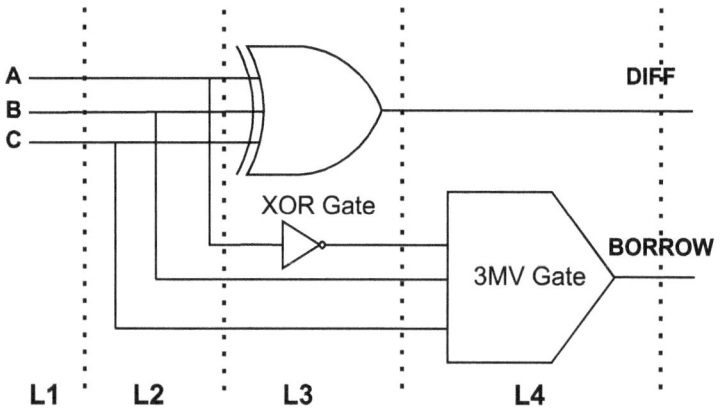

FIGURE 3.27   Full subtractor divided into 4 levels for PTM calculation.

$$L1 = WireA \oplus WireB \oplus WireC \tag{27}$$

$$L2 = FANA \oplus FANB \oplus FANC \tag{28}$$

$$L3 = XOR3 \oplus NOT \oplus I4 \qquad (29)$$

$$L4 = MAJ3 \oplus I2 \qquad (30)$$

Here, I2 and I4 are the identity matrix having size 2×2, 4×4 respectively. WireA, WireB, and WireC represents PTM of three inputs A, B, and C respectively. FANA, FANB, and FANC are FAN-OUT matrix, each of size 2×4 and defined as FAN-OUT = [1 0 0 0; 0 0 0 1]. XOR3, NOT, and MAJ3 are the PTM for 3-input XOR gate, inverter and 3-input majority gate. The overall PTM of FS is calculated with matrix multiplication of all levels as defined in Eq. 31. Utilizing MATLAB, the computation is done and reliability of the circuit is found for varying error rates as shown in Figure 3.28.

$$PTM_{FS} = L1 * L2 * L3 * L4 \qquad (31)$$

3.4.5 COMPUTATION FOR 2×1 MULTIPLEXER

To compute reliability of proposed 2×1 MUX, the circuit is divided into sub levels designated as L1, L2, L3, and L4 as illustrated in Figure 3.29. Using Kronecker

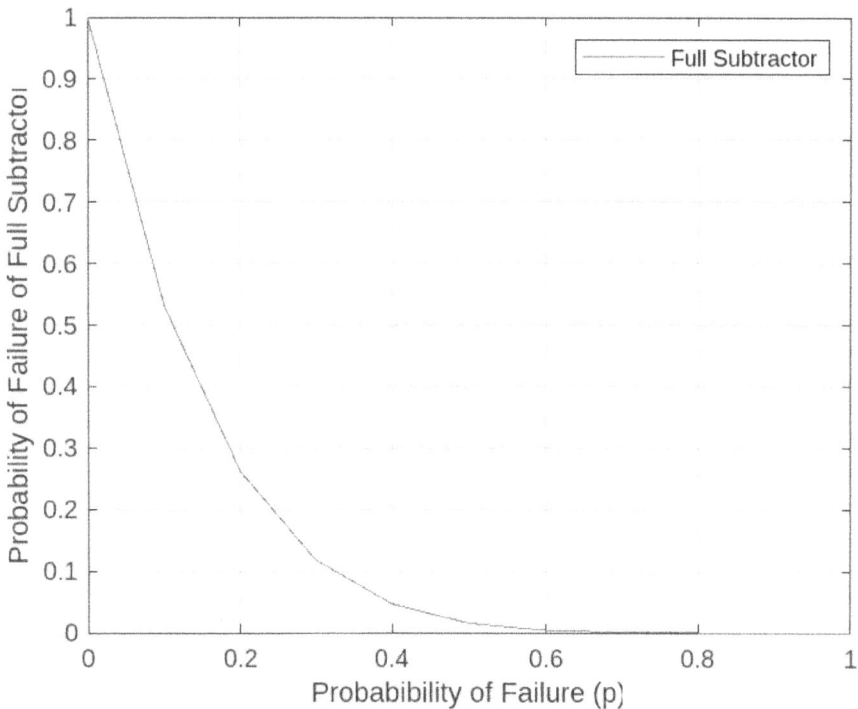

FIGURE 3.28 Reliability of full subtractor with respect to probability of failure.

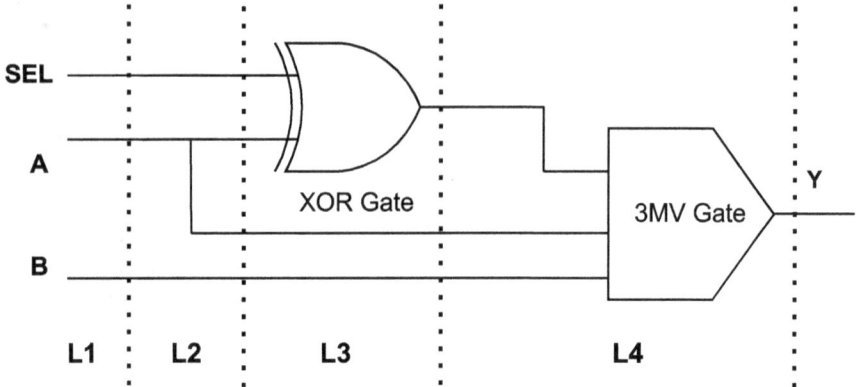

FIGURE 3.29 2 × 1 MUX divided into 4 levels for PTM calculation.

product and matrix multiplication the PTM of individual levels is calculated as per Eq. 32–35.

$$L1 = WireA \oplus WireB \oplus WireC \qquad (32)$$

$$L2 = FANB \oplus I4 \qquad (33)$$

$$L3 = XOR2 \oplus I4 \qquad (34)$$

$$L4 = MAJ3 \qquad (35)$$

Here, I4 represents the identity matrix having size 4×4. WireA, WireB, and WireC represents PTM of three inputs SEL, A, and B respectively. FANB is the FAN-OUT matrix of input B with size 2×4 and defined as FAN-OUT = [1 0 0 0; 0 0 0 1]. XOR2 and MAJ3 are the PTM for 3-input XOR gate and 3-input majority gate. The overall PTM of 2×1 MUX is calculated with matrix multiplication of all levels as defined in Eq. 36. Utilizing MATLAB, the computation is done and reliability of the circuit is found for varying error rates as shown in Figure 3.30.

$$PTM_{FS} = L1 * L2 * L3 * L4 \qquad (36)$$

3.5   CONCLUSION

The proposed set of adders, subtractors and multiplexer in this chapter are efficient in terms of cell count, area, delay, and cost-function. The comparison with prior work has been provided for each combinational circuit. The implementation is done in QCA designer 2.0.1 and power estimation has been performed with QCA Designer-E. The circuits are built using cell interaction approach, therefore the design complexity is highly reduced. The calculations also show a large decrement in cost-function of the

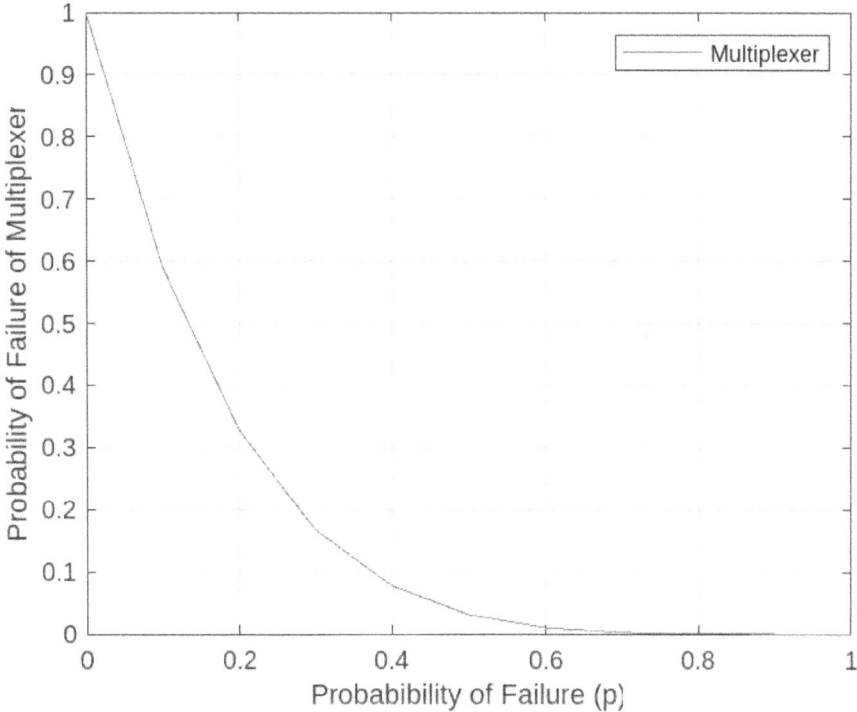

FIGURE 3.30   Reliability of 2 × 1 MUX with respect to probability of failure.

proposed architectures since no inverter and crossover used. The probability matrix computation shows the overall reliability of designs over different values of error rate. The resultant graph shows that the probability of failure of proposed gate is higher for error rates having value greater than 0.6. The designs can be utilized in future to design ripple carry adders, control adder/subtractor, comparator, etc. In addition, multiplexer design can be used in designing of reduced flip-flops, memory and counter in future.

REFERENCES

D. Abedi, G. Jaberipur, and M. Sangsefidi, "Coplanar full adder in quantum-dot cellular automata via clock-zone-based crossover," *IEEE Transactions on Nanotechnology*, vol. 14, no. 3, pp. 497–504, 2015.

P. Z. Ahmad, F. Ahmad, and H. A. Khan, "A new f-shaped xor gate and its implementations as novel adder circuits based quantum-dot cellular automata (QCA)," *IOSR Journal of Computer Engineering (IOSR-JCE)*, vol. 16, no. 3, pp. 110–117, 2014.

P. Z. Ahmad, S. M. K. Quadri, S. M. Tantary, G. M. Wani, F. Ahmad, and A. N. Bahar, "Design of novel QCA-based half/full subtractors," *Nanomaterials and Energy*, vol. 6, no. 2, pp. 59–66, 2017.

D. Ajitha, K. V. Ramanaiah, and V. Sumalatha, "An efficient design of xor gate and its applications using QCA," *i-Manager's Journal on Electronics Engineering*, vol. 5, no. 3, p. 22, 2015.

M. A. Amiri, M. Mahdavi, and S. Mirzakuchaki, "QCA implementation of a mux-based fpga clb," in *2008 International Conference on Nanoscience and Nanotechnology*. IEEE, 2008, pp. 141–144.

S. Angizi, E. Alkaldy, N. Bagherzadeh, and K. Navi, "Novel robust single layer wire crossing approach for exclusive or sum of products logic design with quantum-dot cellular automata," *Journal of Low Power Electronics*, vol. 10, no. 2, pp. 259–271, 2014.

M. Askari and M. Taghizadeh, "Logic circuit design in nano-scale using quantum-dot cellular automata," *European Journal of Scientific Research*, vol. 48, no. 3, pp. 516–526, 2011.

B. Bishnoi, M. Giridhar, B. Ghosh, and M. Nagaraju, "Ripple carry adder using five input majority gates," in *2012 IEEE International Conference on Electron Devices and Solid State Circuit (EDSSC)*. IEEE, 2012, pp. 1–4.

H. Dallaki and M. Mehran, "Novel subtractor design based on quantum-dot cellular automata (qca) nanotechnology," *International Journal of Nanoscience and Nanotechnology*, vol. 11, no. 4, pp. 257–262, 2015.

J. C. Das and D. De, "Shannons expansion theorem-based multiplexer synthesis using QCA," *Nanomaterials and Energy*, vol. 5, no. 1, pp. 53–60, 2016.

T. J. Dysart and P. M. Kogge, "Probabilistic analysis of a quantum-dot cellular automata multiplier implemented in different technologies," in *Proceedings of the 4th Workshop on Non-Silicon Computing*, vol. 46. Citeseer, 2007, pp. 4373–4375.

H. M. Gaur and A. Singh, "Design of reversible circuits with high testability," *Electronics Letters*, vol. 52, no. 13, pp. 1102–1104, 2016.

H. M. Gaur, A. K. Singh, and U. Ghanekar, "A review on online testability for reversible logic," *Procedia Computer Science*, vol. 70, pp. 384–391, 2015.

H. M. Gaur, A. K. Singh, and U. Ghanekar, "Design of reversible arithmetic logic unit with built-in testability," *IEEE Design & Test*, vol. 36, no. 5, pp. 54–61, 2019.

H. M. Gaur, T. Sasamal, A. Singh, A. Mohan, and D. Pradhan, "Reversible logic: An introduction," in: Ashutosh Kumar Singh, Masahiro Fujita, and Anand Mohan (eds) *Design and Testing of Reversible Logic*. Springer, 2020, pp. 3–18.

S. Hashemi and K. Navi, "A novel robust qca full-adder," *Procedia Materials Science*, vol. 11, pp. 376–380, 2015.

M. Hayati and A. Rezaei, "Design of novel efficient adder and subtractor for quantum-dot cellular automata," *International Journal of Circuit Theory and Applications*, vol. 43, no. 10, pp. 1446–1454, 2015.

H. S. Jagarlamudi, M. Saha, and P. K. Jagarlamudi, "Quantum dot cellular automata based effective design of combinational and sequential logical structures," *World Academy of Science, Engineering and Technology*, vol. 60, pp. 671–675, 2011.

V. Jain, D. K. Sharma, and H. M. Gaur, "Area and energy optimized multilayer QCA-based 4n-bit scalable multiplier (m4n–mul)," *The European Physical Journal Plus*, vol. 137, no. 11, p. 1281, 2022.

R. Jaiswal and T. N. Sasamal, "Efficient design of full adder and subtractor using 5-input majority gate in QCA," in *2017 Tenth International Conference on Contemporary Computing (IC3)*. IEEE, 2017, pp. 1–6.

M. Kianpour and R. Sabbaghi–Nadooshan, "Optimized design of multiplexor by quantum-dot cellularautomata," *International Journal of Nanoscience and Nanotechnology*, vol. 9, no. 1, pp. 15–24, 2013.

M. Kianpour, R. Sabbaghi-Nadooshan, and K. Navi, "A novel design of 8-bit adder/subtractor by quantum-dot cellular automata," *Journal of Computer and System Sciences*, vol. 80, no. 7, pp. 1404–1414, 2014.

K. Kim, K. Wu, and R. Karri, "The robust qca adder designs using composable QCA building blocks," *IEEE Transactions on Computer-Aided Design of Integrated Circuits and Systems*, vol. 26, no. 1, pp. 176–183, 2006.

C. Labrado and H. Thapliyal, "Design of adder and subtractor circuits in majority logic-based field-coupled QCA nanocomputing," *Electronics Letters*, vol. 52, no. 6, pp. 464–466, 2016.

S. K. Lakshmi and G. Athisha, "Design and analysis of adders using nanotechnology based quantum dot cellular automata," *Journal of Computer Science*, vol. 7, pp. 1072–1079, 2011.

S. K. Lakshmi, G. Athisha, M. Karthikeyan, and C. Ganesh, "Design of subtractor using nanotechnology based QCA," in *2010 International Conference on Communication Control and Computing Technologies*. IEEE, 2010, pp. 384–388.

C. S. Lent and P. D. Tougaw, "A device architecture for computing with quantum dots," *Proceedings of the IEEE*, vol. 85, no. 4, pp. 541–557, 1997.

C. S. Lent, P. D. Tougaw, W. Porod, and G. H. Bernstein, "Quantum cellular automata," *Nanotechnology*, 4 (1993) 49–57.

W. Liu, L. Lu, M. ONeill, and E. E. Swartzlander, "A first step toward cost functions for quantum-dot cellular automata designs," *IEEE Transactions on Nanotechnology*, vol. 13, no. 3, pp. 476–487, 2014.

C. Mack, "The multiple lives of moore's law," *IEEE Spectrum*, vol. 52, no. 4, pp. 31–31, 2015.

V. A. Mardiris and I. G. Karafyllidis, "Design and simulation of modular 2n to 1 quantum-dot cellular automata (QCA) multiplexers," *International Journal of Circuit Theory and Applications*, vol. 38, no. 8, pp. 771–785, 2010.

D. Mukhopadhyay and P. Dutta, "Quantum cellular automata based novel unit 2: 1 multiplexer," *International Journal of Computer Applications*, vol. 43, no. 2, pp. 22–25, 2012.

M. Poorhosseini and A. R. Hejazi, "A fault-tolerant and efficient xor structure for modular design of complex qca circuits," *Journal of Circuits, Systems and Computers*, vol. 27, no. 07, p. 1850115, 2018.

J. I. Reshi and M. T. Banday, "Efficient design of nano scale adder and subtractor circuits using quantum dot cellular automata," *3rd International Conference on Electrical, Electronics, Engineering Trends, Communication, Optimization and Sciences (EEECOS 2016)*, Tadepalligudem, pp. 1–6, 2016. doi: 10.1049/cp.2016.1508.

A. Roohi, H. Khademolhosseini, S. Sayedsalehi, and K. Navi, "A novel architecture for quantum-dot cellular automata multiplexer," *International Journal of Computer Science Issues*, vol. 8, no. 1, pp. 55–60, 2011.

S. Santra and U. Roy, "Design and implementation of quantum cellular automata based novel adder circuits," *International Journal of Nuclear and Quantum Engineering*, vol. 8, no. 1, pp. 178–183, 2014.

T. N. Sasamal, A. K. Singh, and A. Mohan, "An efficient design of quantum-dot cellular automata based 5-input majority gate with power analysis," *Microprocessors and Microsystems*, vol. 59, pp. 103–117, 2018.

T. Sasamal, H. Gaur, A. Singh, and A. Mohan, "Novel approaches for designing reversible counters," in: Ashutosh Kumar Singh Masahiro Fujita, and Anand Mohan (eds) *Design and Testing of Reversible Logic*. Springer, 2020, pp. 37–48.

T. N. Sasamal, A. K. Singh, and A. Mohan, "An optimal design of full adder based on 5-input majority gate in coplanar quantum-dot cellular automata," *Optik*, vol. 127, no. 20, pp. 8576–8591, 2016.

S. Singh, S. Pandey, and S. Wairya, "Modular design of $2^{\wedge}$ n: 1 quantum dot cellular automata multiplexers and its application, via clock zone based crossover," *International Journal of Modern Education and Computer Science*, vol. 8, no. 7, p. 41, 2016.

F. S. Torres, R. Wille, P. Niemann, and R. Drechsler, "An energy-aware model for the logic synthesis of quantum-dot cellular automata," *IEEE Transactions on Computer-Aided Design of Integrated Circuits and Systems*, vol. 37, no. 12, pp. 3031–3041, 2018.

P. D. Tougaw and C. S. Lent, "Dynamic behavior of quantum cellular automata," *Journal of Applied Physics*, vol. 80, no. 8, pp. 4722–4736, 1996.

P. D. Tougaw and C. S. Lent, "Logical devices implemented using quantum cellular automata," *Journal of Applied Physics*, vol. 75, no. 3, pp. 1818–1825, 1994.

A. Vetteth, K. Walus, V. S. Dimitrov, and G. A. Jullien, "Quantum-dot cellular automata carry-look-ahead adder and barrel shifter," in *IEEE Emerging Telecommunications Technologies Conference*, 2002, pp. 2–4.

W. Wang, K. Walus, and G. A. Jullien, "Quantum-dot cellular automata adders," in *2003 Third IEEE Conference on Nanotechnology, 2003. IEEE-NANO 2003*, vol. 1. IEEE, 2003, pp. 461–464.

4 High-Speed Comparator and Parity Generator Towards Simplified Clocking Circuit in QCA Technology

Trailokya Nath Sasamal, Hari Mohan Gaur,
Vineet Jaiswal, Gaurav Saini and
Ashutosh Kumar Singh

4.1 INTRODUCTION

In the CMOS-based technique, devices with reduced component sizes exhibit higher leakage currents and are more susceptible to noise [1, 2]. Therefore, possible alternatives should be explored to overcome the constraints of current VLSI technology. Diverse nanocomputing devices have been analyzed recently, among which the Quantum-Dot Cellular Automaton (QCA) is less explored in nanotechnology [3, 4]. QCA provides important advantages not available with existing CMOS technology, likewise very low power consumption, high device density, and operating frequency in the THz range [5, 6]. In QCA, all relevant data is expressed as charge reconfiguration. No current flows in QCA-based circuits as information is passed between the diagonal adjacent dots using the polarization principle. Various QCA-based structures have been explored in recent years [7–27], likewise multiplexers, multipliers, adders, ALU, etc.

This book chapter explores the non-majority-based XOR gates for realizing two new QCA architectures. The optimal XOR gate requires a smaller number of cells compared to the existing AND-OR-INVERTER design. This work concentrates on efficient executions of complex designs considering XOR gates as basic building blocks. An ultra-fast, less-complex QCA parity generator and comparator are demonstrated using non-majority-based XOR gates.

This chapter is arranged using the following sections: an overview of QCA basics, XOR gates, and comparators are explored in Section 2. Section 3 illustrates the non-majority-based single-layer architectures of parity generators and comparators based on the optimal XOR structures. Simulation results and associated analyses are presented in Section 4, considering previous analysis, and in Section 5, we conclude the chapter.

DOI: 10.1201/9781003361633-4

4.2 BACKGROUND

4.2.1 QCA BASICS

In QCA, each cell consists of four quantum dots and two unrestricted electrons. The two free electrons move within diagonal adjacent cells, and four dots control it. These two electrons settle at diagonally opposite positions in the quantum cell due to the repulsion between the electrons. There are two stable polarization states based on the electrons' position [1]. The two possible polarization states with tunnel junctions are shown in Figure 4.1(a). Electrons tunnel through the barriers between the dots when a suitable voltage is applied. The logic values are transmitted through cells due to interactions between cells and polarization of neighboring cells. Figure 4.1 demonstrates basic QCA elements like the majority voter (MV) gate and NOT gate. A collection of side-by-side QCA dots implements a QCA line that enables computation-free data transmission between two terminals, as demonstrated in Figure 4.1(b). Figure 4.1(c) illustrates two distinct forms of inverters. Two different forms of the majority gate (MG) are shown in Figure 4.1(d). AND logic/OR logic is implemented by setting one of the cell values to be '0'/'1'.

All QCA circuits must be clocked appropriately to provide the power required to drive the QCA cells. Logic value can propagate between the cells using the proper clocking. The number of cells is distributed according to the number of clock zones. Each clock zone is independent, and clock signals are phase shifted by 90° with the adjacent zone [2]. Clock zones help the logic value sequencing and its computation. Most researchers used four clock zones to drive input data to the desired output.

4.2.2 XOR STRUCTURES

XOR gates play an important role in designing different digital systems, likewise ALUs, parity checkers and generators, comparators, etc. Different designs of QCA-based XOR are described below. These layout designs are generally categorized into the following types:

1. Using 3-i/p majority gate (MV3).
2. Using 3-i/p and 5-i/p majority gates.
3. Using explicit cell interactions based on non-majority structures.

Figure 4.2(a) demonstrates the schematic design of the first type of design using three i/p majority gates and one inverter logic.

$$\mathrm{MAJ3}\Big(\big(\mathrm{MAJ3}\big(A,B,0\big)\big),\mathrm{MAJ3}\big(A,B,1\big),1\Big) = \mathrm{MAJ3}(\bar{A}B, A\bar{B},1)$$

$$= \bar{A}B + A\bar{B} \tag{1}$$

The Boolean expression of the first architecture is expressed as in equation (1):

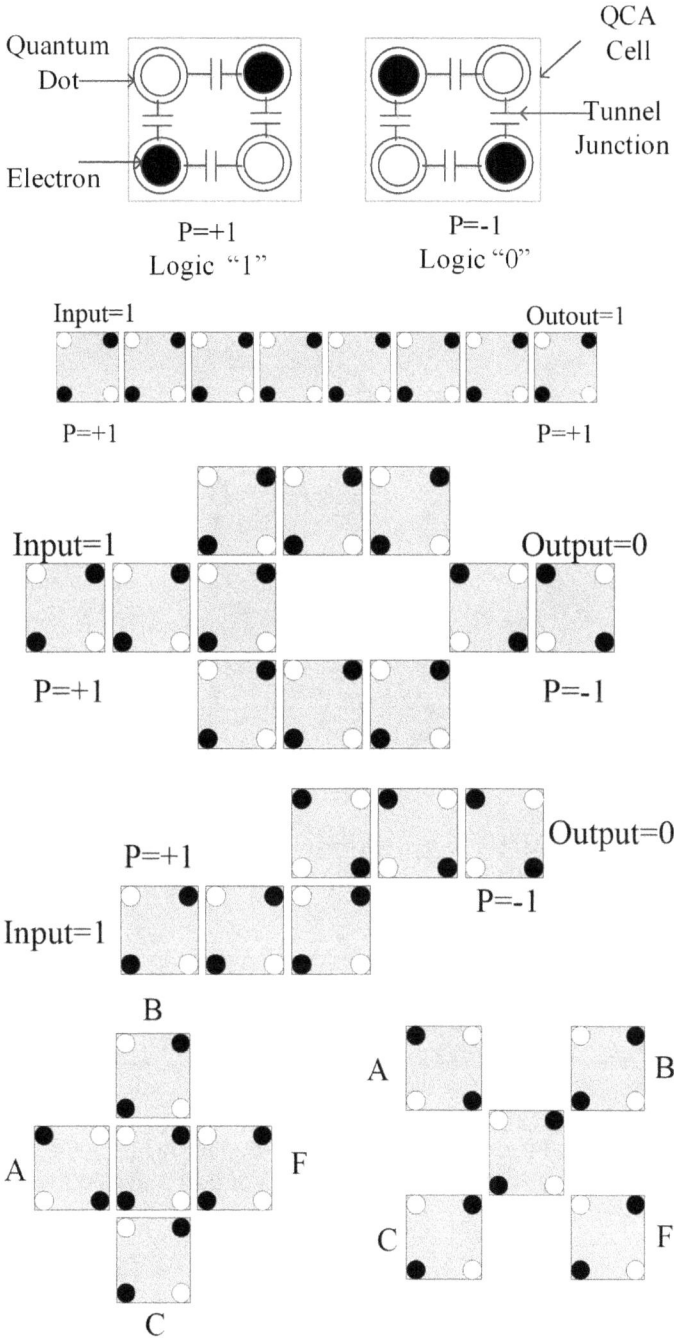

FIGURE 4.1 QCA (a) QCA's different polarization state (b) Wire architecture (c) Realization of two distinct NOT gate structures (d) Layout design of majority voter.

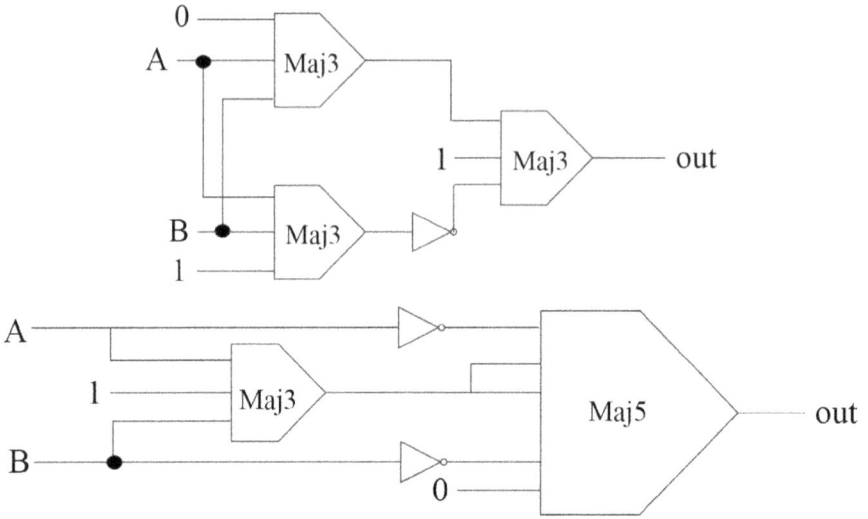

FIGURE 4.2 Conventional XOR gate schematic design (a) Using 3-i/p MV (b) Using 5-i/p and 3-i/p MV.

Figure 4.2(b) demonstrates a schematic design for the second type of architecture. This architecture uses two inverter logic gates, a 5-i/p majority gate, and one 3-i/p majority gate.

$$\begin{aligned} &\text{MAJ5}\big(\text{MAJ3}\big(A,B,1\big),\text{MAJ3}\big(A,B,1\big),\bar{A},\bar{B},0\big) \\ &= \text{MAJ5}\big(\big(A+B\big),\big(A+B\big),\bar{A},\bar{B},0\big) = \bar{A}B + A\bar{B} \end{aligned} \tag{2}$$

The Boolean expression of the second architecture can be expressed as in equation (2):

The next type of XOR gate is implemented utilizing the non-majority gate and explicit interactions of the cells. Therefore, the conventional method of utilizing MGs is not used to implement XOR gate designs. In this chapter, we used one non-majority gate structures of two-input XOR gates that is introduced in [17]. The explicit interactions of the cells achieve the implementation of these structures. In this way, the output is affected by cells within the effective radius. Figure 4.3 shows the XOR gate structure. It requires 13 cells, covers an area of 0.01 μm², and has a 0.25 clock cycle propagation delay [17].

4.2.3 COMPARATOR DESIGNS

The comparator is the fundamental computational component in microcontrollers and CPUs. A comparator architecture takes two logic values as inputs, compares those two values, and produces three possible results, i.e. (A=B, A<B, A>B). Numerous architectures of single-bit magnitude comparators have been described

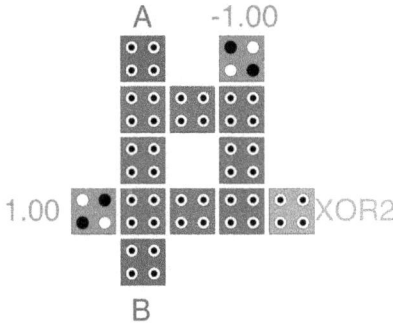

FIGURE 4.3   XOR gate optimal designs in [17].

[11–21]. [17–19, 21] utilizes XOR gates, MGs, and inverter logic in all structures. Whereas layout designs by [11–16, 20] require MGs and inverter logic to implement comparator.

The polarization of the output cell is $8.56 \times e^{-001}$, compromising signal veracity in huge and difficult designs [20]. In the QCA architecture, i/p B is not accessible, so input cell B must be rearranged in the cascade architecture. Furthermore, accessing the A=B (one of the o/p) requires a multilayer crossing for fan-out structures. This leads to additional QCA cells, and the cell count exceeds 47 cells. The design described in [19] requires a multilayer or single-layer crossing network to access the i/p (B) in multi-bit comparators or further complex designs that require the previous stage's output to be connected to the B input. Therefore, according to the authors' report, the total number of cells exceeds 44. So, the design's occupied area also increases. The design in [21] also faces the same problem and requires an extra cell to access the o/p cell. Thus, the implemented design needs reorganized QCA cells, consequential in a total number exceeding 40. Wang et al. [17] presented an ideal design of a single-bit comparator. This design solves all the above-discussed problems, but after observing the design thoroughly, A < B (o/p logic) needs to be corrected. Therefore, adding an extra MG is possible to update these designs. The consequences of this is a rise in cell counts and occupied area over those discussed earlier.

4.3   PROPOSED QCA DESIGNS

We implemented a single-bit comparator and complex parity generator layout designs with the help of the above to show the applicability of the non-majority-based XOR design.

4.3.1   SINGLE-BIT COMPARATOR DESIGN

This section described a new QCA layout design for a single-bit magnitude comparator. By reducing the number of crossovers, we implement a design that helps to propose big and scalable layout architectures. The Boolean expressions for outputs ALB, AEB, and AGB are demonstrated below using A and B as inputs.

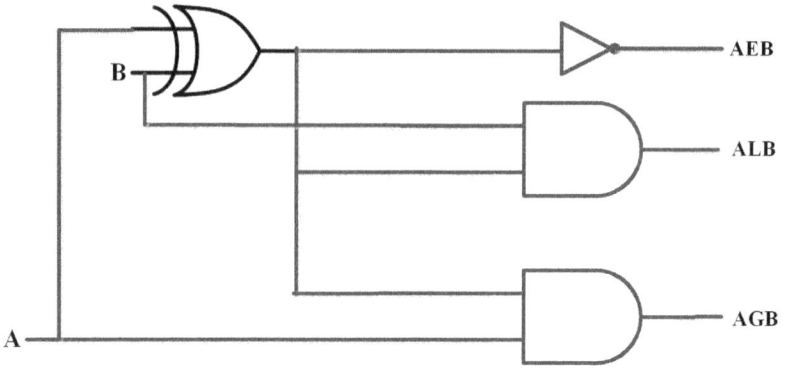

FIGURE 4.4   Single-bit comparator's schematic design.

$$ALB = (A \oplus B).B = \bar{A}B$$

$$AEB = (\overline{A \oplus B}) = A \odot B$$

$$AGB = (A \oplus B).A = A\bar{B} \tag{3}$$

Figure 4.4 demonstrates the schematic design of a single-bit comparator. The architectural design can be implemented using two majority gates (Maj3), two inverters, and one XOR gate.

Figure 4.5 shows the layout design of the implemented single-bit coplanar QCA-based comparator. The advantage of this design is that all input and output ports are easy to access on a single layer without using a multilayer structure. This layout can be implemented using two inverter logic, two three-input majority gates, and one optimal XOR gate [17], presented by a red square box. The cell complexity or the number of cells also decreases, which is 38 QCA cells. The total area occupancy is 0.04 μm², and the area coverage of the QCA cells is 0.012 μm², which is 30% of the total area. This new coplanar design of the comparator allows data from the input port to the output port, and it needs only four clock phases to produce a correct output. So, the total delay of the proposed comparator is equivalent to a single clock cycle.

4.3.2 PARITY GENERATOR CIRCUIT

The essential element of digital systems is the parity generator. It is mostly used for error detection techniques. This section introduces a new QCA layout design for the 4-bit even parity generator. The optimal design developed in [9] implements a parity generator structure. Here the number of crossovers is less, which is helpful in large and complex designs. The logical expression for output P of the parity generator using A, B, C, and D as inputs:

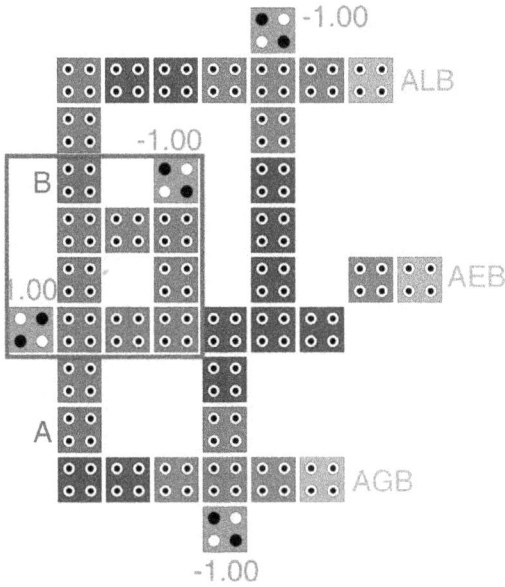

FIGURE 4.5 Novel single-bit comparator using optimal XOR gate.

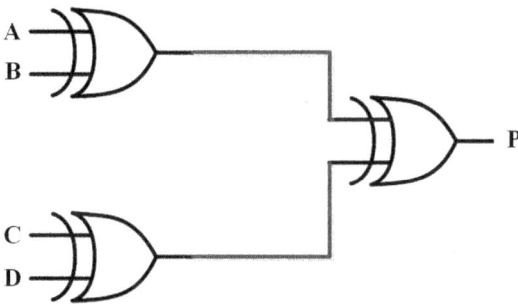

FIGURE 4.6 Depicts the schematic design of a 4-bit parity generator. This design can be implemented using three XOR gates.

$$P = \left(\left(A \oplus B\right) \oplus \left(C \oplus D\right)\right) \tag{4}$$

Figure 4.6 depicts a 4-bit parity generator's schematic design (even parity).

The QCA layout design of the implemented four-bit coplanar parity generator is demonstrated in Figure 4.7. The advantage of the single-layer (without crossover) structure is that all input and output cells are easy to access. This layout design needs three optimal XOR gates which were implemented by Wang et al. [17] and it is shown using a red square box. There is a need for two clock phases to produce

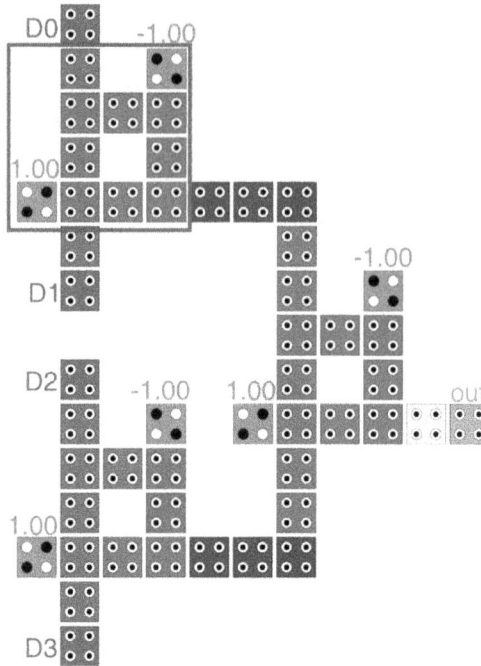

FIGURE 4.7   Proposed QCA layout design of the 4-bit parity generator.

each XOR gate output. Also, the cell complexity or the number of cells is 50 because we require 50 QCA cells for implementation. The total area occupancy is $0.08\ \mu m^2$, and the area coverage of the QCA cells is $0.016\ \mu m^2$, which is 20% of the total area. The proposed compact coplanar design of the parity generator requires four clock zones to transmit logic value from the i/p pin to the o/p pin to produce the correct output. Therefore, the total delay of the proposed 4-bit even parity generator is only one clock cycle. In this research, we also developed 8-bit and 16-bit parity generators considering the optimal XOR gates, which are demonstrated in Figure 4.8 and Figure 4.11 respectively.

4.4   SIMULATION RESULTS AND ANALYSIS

Proposed layout designs are functionally verified after implementation and simulation analysis. All the layout designing and simulation analysis is done on QCA Designer 2.03 [28]. The simulation is done using a Bi-stable approximation engine and its default settings.

The simulation waveform of the presented 1-bit comparator is shown in Figure 4.9. It verifies the correctness of the proposed structure. For simplicity we have shown the clock 2 as the comparator delivers correct result after two clock phases, i.e. at the clock 2, which is shown by red box.

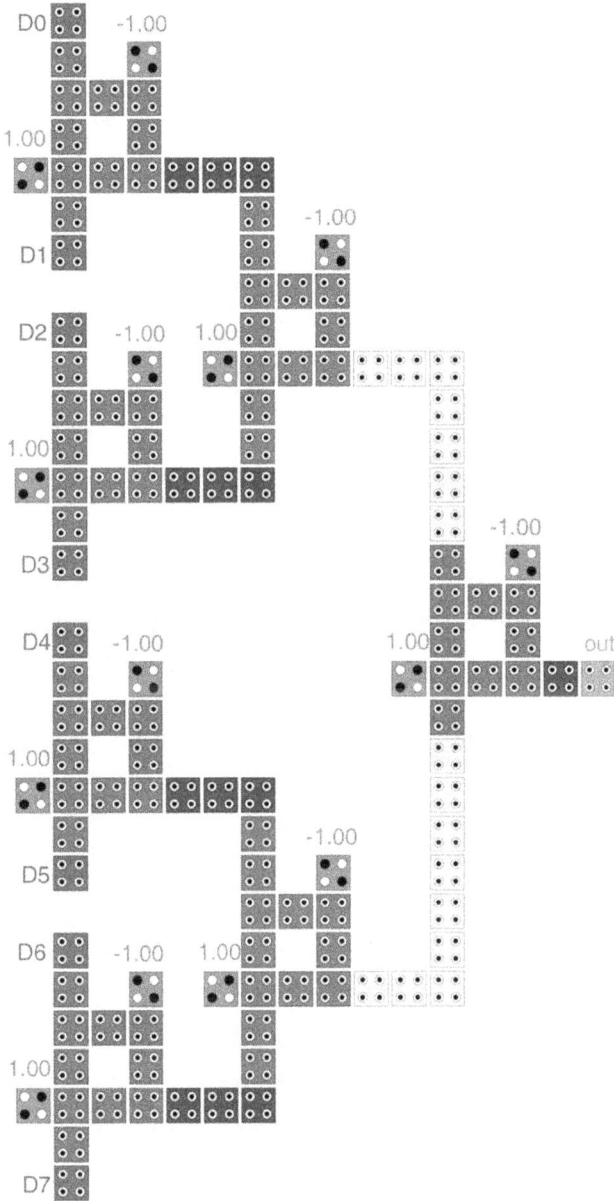

FIGURE 4.8 Proposed QCA layout design of the 8-bit parity generator.

The simulation waveform of the presented four-bit parity generator is shown in Figure 4.10. It verifies the accuracy of the proposed structure. For simplicity, we have shown the clock 3 as the parity generator delivers correct result after three clock phases, i.e. at the clock 3, which is shown by black box. Table 4.1 summarizes the proposed comparator design and the previous architectures taking different characteristics

Simulation Results

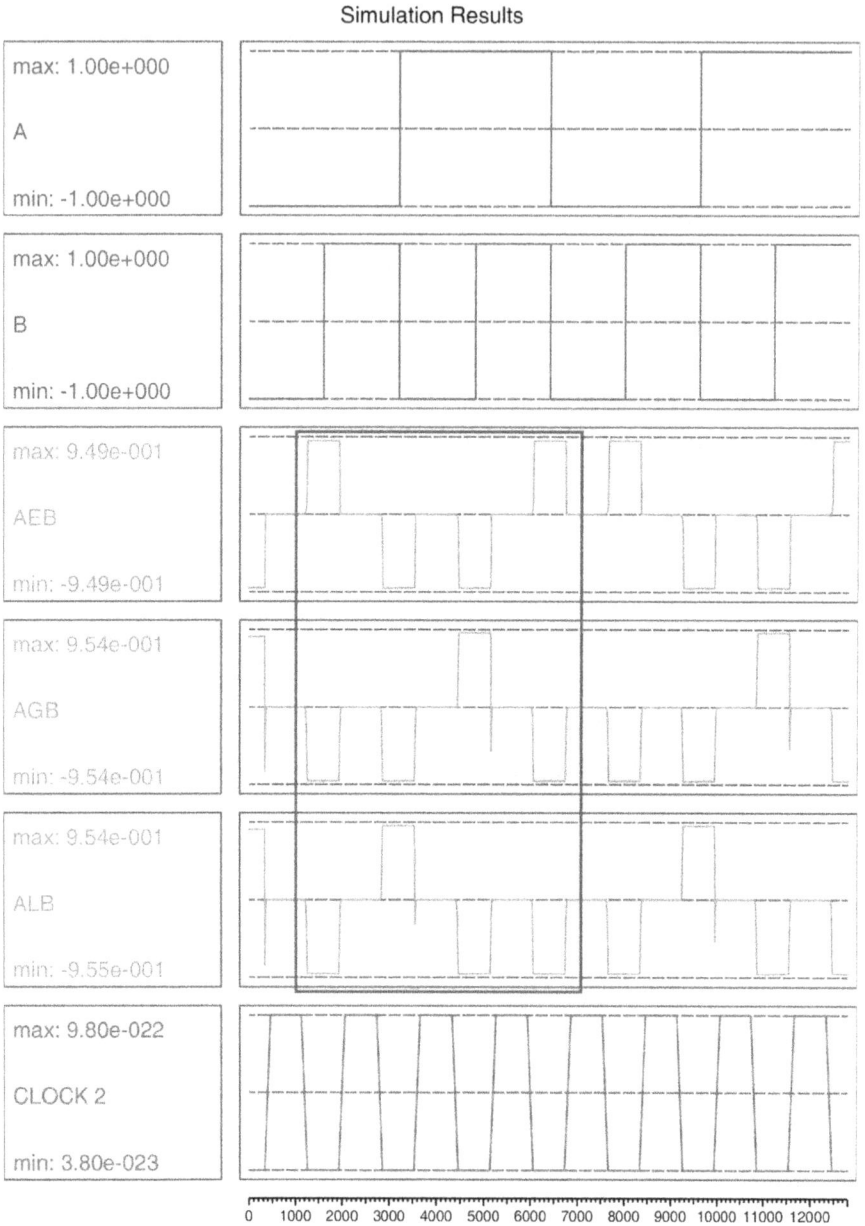

FIGURE 4.9 Output waveform of the presented single-bit comparator.

parameters. These parameters include efficient complexity, area utilize, number of QCA cells, and delay.

$$\text{Efficient complexity} = (\text{Cell} \times \text{Area}^{1/n}) \tag{5}$$

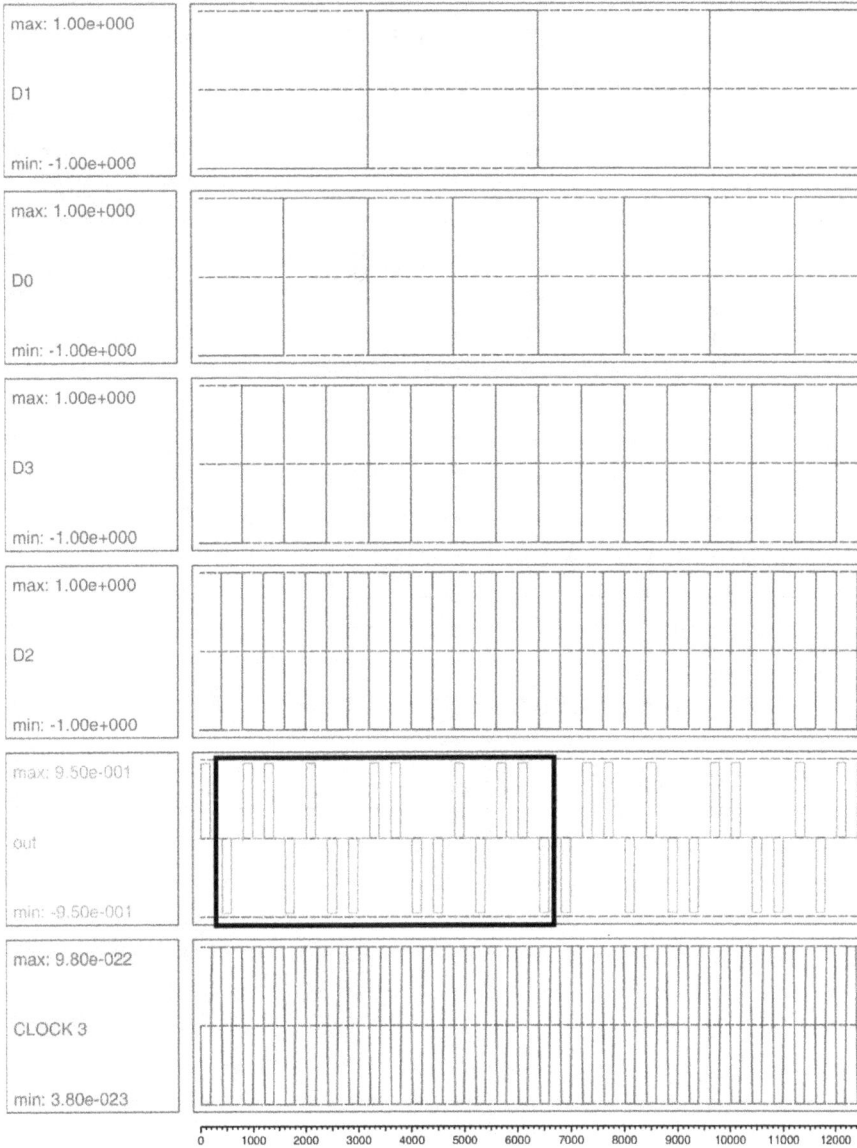

FIGURE 4.10   Output waveform of the presented four-bit parity generator.

Here n = no. of layers. It is worth noting that the presented comparator layout outperforms many available structures [11–20] in terms of common design parameters. It is noted that design in [21] is close to the proposed design but quick observation reveals that design in [21] is less flexible and requires additional cells and crossover to facilitate input/output accessible to other units in complex designs.

TABLE 4.1

Comparative analysis of presented comparator and existing designs

Design	Number of Cells	Total Layout area (µm²)	Cell Area(µm²)	Area utilize (%)	EC	Latency	Crossover
[11]	117	0.182	0.037	20.56	21.29	2	Single
[12]	85	0.064	0.027	42.18	5.44	1.5	Single
[13]	100	0.110	0.032	29.09	47.91	0.05	Multilayer
[14]	79	0.032	0.026	81.25	25.08	1	Multilayer
[15]	81	0.06	-	-	33.38	0.75	Multilayer
[16]	73	0.053	0.024	44.28	28.58	1	Multilayer
[17]	47	0.036	0.015	41.67	1.69	0.75	Single*
[18]	54	0.04	-	-	18.46	1	Multilayer
[19]	44	0.038	0.014	36.84	1.67	1	Single*
[20]	47	0.042	0.015	35.71	1.97	0.5	Single*
[21]	40	0.05	0.012	24	2	0.75	Single*
Proposed	40	0.06	0.012	30	1.52	0.75	Single

* Designs require extra crossover and cells

TABLE 4.2

Comparative analysis of parity generators of varying length

Design	No. Bits	Cell Count	Delay (Clock phases)	QCA Layout Area (µm²)
[22]	4	187	2.75	0.32
	8	456	4	0.92
	16	1024	5.25	2.41
[23]	4	188	2.25	0.2
	8	369	2.25	0.49
	16	847	3.25	1.46
[24]	4	111	2	0.14
	8	269	3	0.43
	16	603	4	1.13
[25]	4	98	2	0.11
	8	241	3	0.37
	16	537	4	1.04
[26]	4	97	1.75	0.1
	8	235	2.75	0.3
	16	523	3.75	0.76
[27]	4	87	1.75	0.10
	8	213	2.75	0.30
	16	480	3.75	0.81
Proposed	4	50	1	0.08
	8	126	1.5	0.22
	16	294	2.25	0.58

Table 4.2 summarizes the proposed even parity generator design and the previous architectures taking different characteristics parameters considering varying lengths (4, 8, 16 bits). These parameters include efficient complexity, area utilize, number of QCA cells, and delay. As a result, the proposed designs offer significant improvements over prior comparable designs [22–27] in terms of common design parameters. For example, 16-bit parity generator is illustrated in Figure 4.11. This

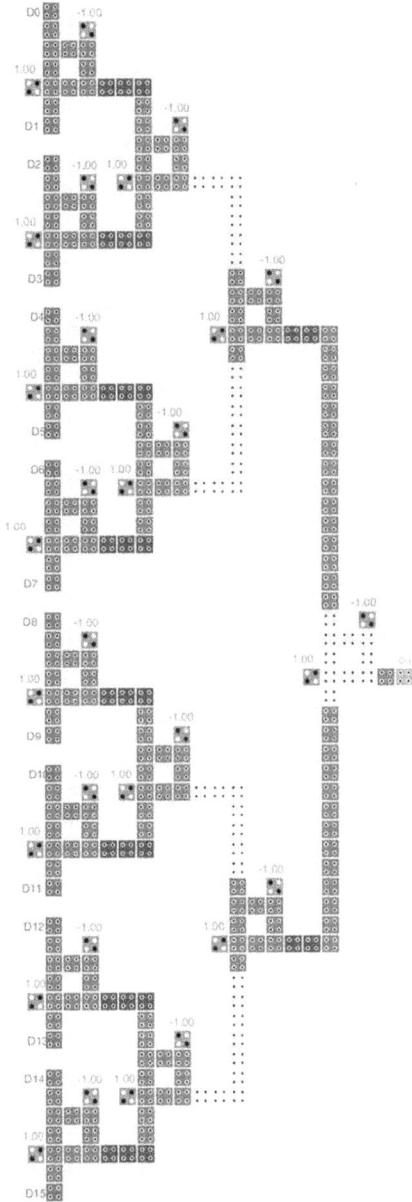

FIGURE 4.11 Layout structure of the presented QCA-based sixteen-bit parity generator.

layout requires 294 QCA cells, which is 38% lesser than the design [27]. All the proposed designs encourage any of the existing realistic clocking schemes like 2DD, USE. In other words, the proposed designs can be easily modified according to the existing clocking schemes with additional cells and delay, whereas most existing designs are not following existing clocking schemes because of the design limitations like not flexible in terms of cell placement and routing.

4.5 CONCLUSION

In this chapter, an optimal XOR gate is used to construct two optimal QCA structures. These structures are realized using XOR gates based on the non-majority gate and three input majority gates. Conventional QCA XOR gate is based on three input majority gates whereas the optimal non-majority-based XOR gate is realized on the principle of cell interaction. Non-majority-based XOR gate enables a reduction in total cell area along with delay which facilitates complex QCA layouts. First, we have designed a coplanar one-bit comparator with three outputs. Furthermore, we also designed parity generators of lengths 4, 8, and 16 bits to showcase the advantages of the optimal XOR gate. All the proposed designs encourage any of the existing realistic clocking schemes like 2DD and USE. Comparative analysis shows that proposed designs have superior performance with respect to prior comparator and parity generator designs when compared with common design parameters. In future, these designs can be used to implement complex structures as it facilitates real clocking schemes with optimal area and delay.

REFERENCES

[1] C. S. Lent, P. D. Tougaw, W. Porod, and G. H. Bernstein, "Quantum cellular automata," *Nanotechnology*, vol. 4, pp. 49–57, Jan. 1993.

[2] C. S. Lent, P. D. Tougaw, and W. Porod, "Bistable saturation in coupled quantum dots for quantum cellular automata," *Appl. Phys. Lett.*, vol. 62, no. 7, pp. 714–716, 1993.

[3] G. L. Snider *et al.*, "Quantum-dot cellular automata: Review and recent experiments," *J. Appl. Phys.*, vol. 85, pp. 4283–4285, Apr. 1999.

[4] M. Macucci *et al.*, "A QCA cell in silicon-on-insulator technology: Theory and experiment," *Superlattices Microstruct.*, vol. 34, nos. 3–6, pp. 205–211, Sep.–Dec. 2003.

[5] P. D. Tougaw, C. S. Lent, "Dynamic behavior of quantum cellular automata," *J Appl Phys*, vol. 80, pp. 4722–4736, 1996.https://doi.org/10.1063/1.363455

[6] C. S. Lent, P. D. Tougaw, "A device architecture for computing with quantum dots," *Proc IEEE*, vol. 85, no. 4, pp. 541–557, 1997. https://doi.org/10.1109/5.573740

[7] T. N. Sasamal, A. K. Singh, A. Mohan, *Quantum-Dot Cellular Automata Based Digital Logic Circuits: A Design Perspective*, Vol. 879. Springer Nature, 2019.

[8] T. N. Sasamal, A. K. Singh, U. Ghanekar. "Design of non-restoring binary array divider in majority logic-based QCA," *Electronics Letters*, vol. 52, no. 24, pp. 2001–2003, 2016.

[9] T.N. Sasamal, A.K. Singh, U. Ghanekar, "Efficient design of coplanar ripple carry adder in QCA," *IET Circuits Devices Syst.*, vol. 12, pp. 594–605. 2018. https://doi.org/10.1049/iet-cds.2018.0020

[10] A. Chudasama, T. N. Sasamal, J. Yadav, "An efficient design of Vedic multiplier using ripple carry adder in Quantum-dot Cellular Automata," *Comput. Electr. Eng.*, vol. 65, pp. 527–542, 2018, ISSN 0045-7906,

[11] M. Abdullah, Al Shafi, A. N. Bahar, "Optimized design and performance analysis of novel comparator and full adder in nanoscale," *Cogent Engineering*, vol. 3, no. 1, p. 1237864, 2016.

[12] S. Erniyazov, J. C. Jeon, "Area efficient magnitude comparator based on QCA," *Adv. Sci. Technol. Lett.* vol. 150, pp. 75–79, 2018.

[13] L. Jun-wen, X. Yin-shui, A novel design of quantum-dots cellular automata comparator using five-input majority gate. In: 2018 14th IEEE International Conference on Solid-State and Integrated Circuit Technology (ICSICT), pp. 1–3, 2018.

[14] M. Hayati, A. Rezaei, "Design and optimization of full comparator based on quantum dot cellular automata," *ETRI Journal*, vol. 34, pp. 284–287, 2012.

[15] D. Ajitha, K. V. Ramanaiah, V. Sumalatha, A novel design of cascading serial bit-stream magnitude comparator using QCA, Advances in Electronics, Computers and Communications (ICAECC), International Conference, pp. 1–6, 2014.

[16] B. Ghosh, S. Gupta, S. Kumari, Quantum dot cellular automata magnitude comparators. IEEE Int Conf Electron Dev Solid State Circuit, 1–2, 2012.

[17] L. Wang, G. Xie, "A novel XOR/XNOR structure for modular design of QCA circuits", *IEEE Trans. Circuits Syst. II: Express Br.*, vol. 67, no. 12, pp. 3327–3331, 2020.

[18] H. Thapliyal Roohi, R. F. Demara, "Wire crossing constrained QCA circuit design using bilayer logic decomposition," *Electron. Lett*, vol. 51, pp. 1677–1679, 2015.

[19] A. Khan, R. Arya, "High performance nanocomparator: a quantum dot cellular automata based approach," *J Supercomput.*, vol. 78, pp. 1–17, 2021.

[20] A. N. Bahar, K. Roy, M. Asaduzzaman, M. M. R. Bhuiyan, "Design and implementation of 1-bit comparator in quantum-dot cellular automata (QCA)," *Cumhuriyet. Sci. J.*, vol. 38, pp. 146–152, 2017.

[21] A. H. Majeed, M. S. Zainal, E. Alkaldy, D. Md. Nor, "Single-bit comparator in quantum-dot cellular automata (QCA) technology using novel QCA-XNOR gates," *J. Electron. Sci. Technol*, vol. 19, no. 3, 263–273, 2021, 100078, ISSN 1674-862X, https://doi.org/10.1016/j.jnlest.2020.100078.

[22] M. T. Niemier, *Designing Digital Systems in Quantum Cellular Automata*. University of Notre Dame, Notre Dame, 2004.

[23] S. Angizi, E. Alkaldy, N. Bagherzadeh, K. Navi, "Novel robust single layer wire crossing approach for exclusive or sum of products logic design with quantum-dot cellular automata," *J. Low Power Electron.*, vol. 10, no. 2, pp. 10(2), 259–271, 2014.

[24] M. Poorhosseini, A. R. Hejazi, "A fault-tolerant and efficient XOR structure for modular design of complex QCA," *J. Circuits Syst. Comput*, vol. 27, pp. 1850115, 2018.

[25] S. Sheikhfaal, S. Angizi, S. Sarmadi, M. H. Moaiyeri, S. Sayedsalehi, "Designing efficient QCA logical circuits with power dissipation analysis," *Microelectron. J*, vol. 46, no. 6, pp. 462–471, 2015.

[26] T. N. Sasamal, A. K. Singh, U. Ghanekar, Design and analysis of ultra-low power QCA parity generator circuit, *Advances in Power Systems and Energy Management, Lecture Notes in Electrical Engineering*, Vol. 436. Springer, Singapore, pp. 341–354, 2018.

[27] G. Singh, R. K. Sarin, B. Raj, "A novel robust exclusive-OR function implementation in QCA nanotechnology with energy dissipation analysis," *J. Comput. Electron*, vol. 15, no. 2, pp. 455–465, 2016.

[28] K. Walus, T. J. Dysart, G. A. Jullien, and R. A. Budiman, "QCA Designer: A rapid design and Simulation tool for quantum-dot cellular automata," *IEEE Trans. Nanotechnol.*, vol. 3, no. 1, pp. 26–31, Mar. 2004.

5 Towards Effective Multiplexer Circuit Design in QCA Technology

Abdalhossein Rezai[1], Davood Aliakbari[2],*
Asghar Karimi[3] and Sasan Ansarian Najafabadi[4]
[1] Department of Electrical Engineering, University of Science and Culture, Tehran, Iran, rezai@usc.ac.ir
[2] ACECR institute of higher education, Isfahan branch, Isfahan, aliakbaridavood395@gmail.com
[3] ACECR institute of higher education, Isfahan branch, Isfahan, karimi@jdeihe.ac.ir
[4] ACECR institute of higher education, Isfahan branch, Isfahan, Sasan.ansarian@yahoo.com
*Corresponding author

5.1 INTRODUCTION

Nowadays, the conventional CMOS technology is faced with scaling issues at nano-scale. The Quantum-Dot Cellular Automata (QCA) technology is a novel nanotech-nology that is a candidate to replace CMOS technology in digital circuits design [1, 2]. The QCA technology provides benefits compared to CMOS technology such as high density and high switching frequency. The basic unit in this technology is quantum cell that have two electrons in four dots. The reconfiguration of cell charge has a basic role in this technology. All digital circuits can be designed and implemented at nano-scale using this technology [3–5].

On the other hand, the multiplexer (MUX) circuit is a basic part in circuits design. This digital circuit is used for selecting one input of several inputs and displays it in its output. As a result, it has several applications in digital circuits design such as ALU [3, 6]. Thus, improving the performance of this circuit is one of research focus in this field. Several attempts [6–23] have been reported for efficiency improvement in the MUX circuits.

This chapter presents and evaluates MUX circuits in the QCA technology. An efficient 2:1 QCA MUX circuit is presented, which can be extended for 2^n:1 QCA MUX. The design strategy in the designed 2^n:1 QCA MUX circuits is using a cost-effective architecture and path-planning design, which can decrease the design costs. The QCADesigner tool version 2.0.3 is utilized to simulation the QCA MUX circuits. The results demonstrate that the designed 2^n:1 QCA MUX circuits have 15 (0.01

DOI: 10.1201/9781003361633-5

μm^2), 50 (0.05), and 159 (0.18 μm^2) cells (area) for n = 1, 2 and 3, respectively. The comparison results indicate that our 2^n:1 MUX circuits have advantages compared to other 2^n:1 QCA MUX circuits.

This chapter is presented as follows. The next section provides the QCA technology concepts, and an overview of MUX circuits. Then, the proposed 2^n:1 QCA MUX circuit is presented for n = 1, 2, and 3 in Section 3. The designed circuits are evaluated and compared in Section 4. Finally, this chapter is concluded in Section 5.

5.2   BACKGROUND

This section outlines the concepts of the QCA technology and an overview of the MUX circuit.

5.2.1   QCA CELL

The QCA cell is a component of QCA technology. In this technology, each square-shaped quantum cell has two electrons and four dots. The mathematical modes of electron placement are six, but due to Coulomb's response, there will be no more than two stable states. Digital logic 1 and 0 represent these two stable states. The QCA cell stable stats is displayed in Figure 5.1 [4, 24, 25].

It should be noted that there are two kinds of QCA cells, 45° and 90° cells [7, 26]. Figure 5.2 displays these two types of cells [7, 26].

Based on this figure, these two formats are formed based on the place of dots in cells [7, 26].

5.2.2   QCA GATES

The digital circuits in the QCA technology can be implemented using three fundamental gates, majority gate, inverter gate, and XOR gate [7, 27, 28]. Figure 5.3 displays a simple QCA majority gate [7].

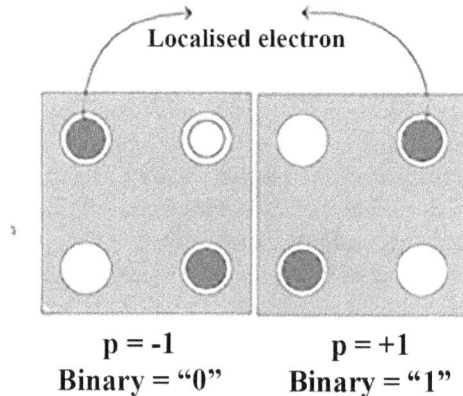

p = -1   **p = +1**
Binary = "0"   **Binary = "1"**

FIGURE 5.1   The QCA cell stable states.

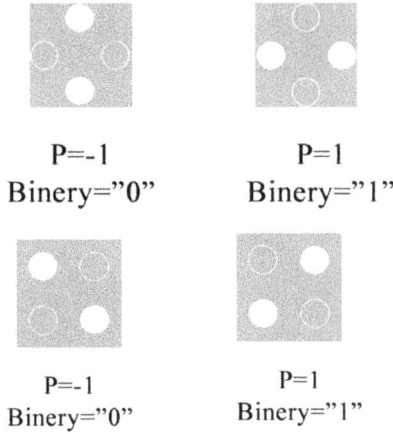

P=-1
Binery="0"

P=1
Binery="1"

P=-1
Binery="0"

P=1
Binery="1"

FIGURE 5.2   The QCA cell (a) 45 ° cell (b) 90 ° cell.

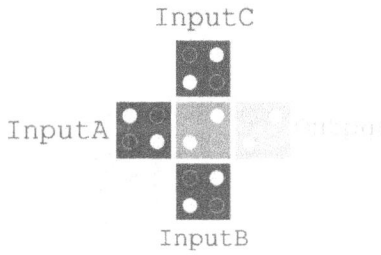

InputC

InputA

InputB

FIGURE 5.3   A simple majority gate.

This QCA gate has even inputs and an output. The output indicates the state of the majority of inputs. So, the output of this QCA gate can be computed as follows for three inputs A, B, and C [1, 7, 29].

$$\text{Output} = BA + CB + AC \tag{1}$$

Based on the utilized QCA cells, the existing majority gate can be categorized in two groups, Rotate Majority Gate (RMG) and Original Majority Gate (OMG). Figure 5.4 displays these two groups of majority gate [7].

Figure 5.5 displays QCA inverter gate [7].

This QCA gate is composed of an input and an output. In this gate, the inverse of input is appeared in the output [7].

The QCA XOR gate is another basic gate in this technology. This gate has n inputs and an output. Figure 5.6 displays the 3-input QCA XOR gates [3].

The output of the QCA XOR gate for three inputs A, B, and C is computed as follows [3].

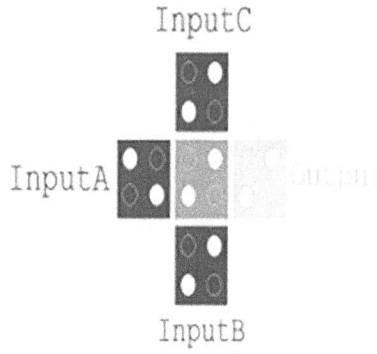

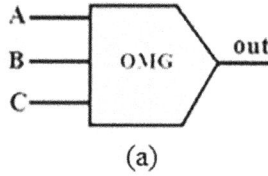

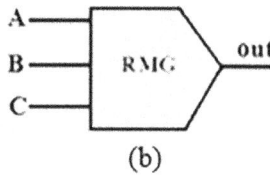

FIGURE 5.4   Two groups of majority gate (a) OMG (b) RMG.

FIGURE 5.5   The QCA inverter gate.

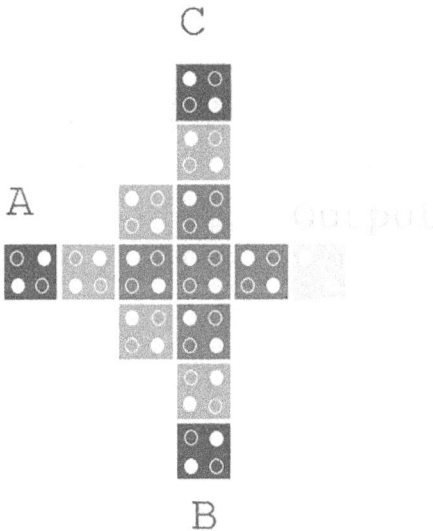

FIGURE 5.6   The QCA XOR gate.

$$Output = XOR3 = A \oplus B \oplus C = \overline{A}\overline{B}C + \overline{A}B\overline{C} + A\overline{C}\overline{B} + ABC \tag{2}$$

5.2.3   MUX Circuits

The MUX circuit is a basic component in circuits design. This circuit can select its output form 2^n inputs by using n address lines. Figure 5.7 displays two frequently used 2:1 MUX circuits [7].

Multiplexer design is in focal point of researches due to the vital role of this circuit in circuits design. As a result, researchers [7–13, 15–19, 22, 23, 30] have attempted to increase the efficiency of this circuit.

Rashidi et al. [7] have developed a coplanar 2:1 QCA MUX circuit. This circuit has 15 cells and 0.01 $\mu m^{2\,area}$. Sen et al. [8] have offered a coplanar 2:1 QCA MUX

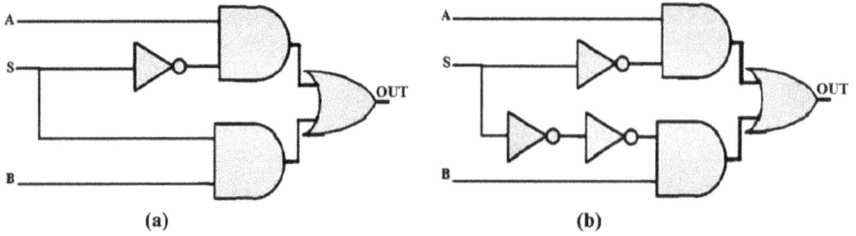

FIGURE 5.7   Two frequently used circuits for 2:1 MUX circuit.

circuit. This circuit has 23 cells and 0.02 $\mu m^{2\,area}$. Hashemi et al. [9] have reported a coplanar 2:1 QCA MUX circuit. This circuit requires 36 cells and 0.06 $\mu m^{2\,area}$. Mardiris et al. [10] have reported a coplanar 2:1 QCA MUX circuit. This circuit requires 67 cells and 0.14 μm^2 area. Mardiris et al. [11] have developed a coplanar 2:1 QCA MUX circuit. This circuit requires 56 cells and 0.07 μm^2 area. Roohi et al. [12] have presented a coplanar 2:1 QCA MUX circuit. This circuit requires 27 cells and 0.03 μm^2 area. Sen et al. [13] have reported a coplanar layer 2:1 QCA MUX circuit. This circuit requires 19 cells and 0.02 μm^2 area. Sen et al. [15] have reported a coplanar 2:1 QCA MUX circuit. This circuit requires 19 cells and 0.02 μm^2 area. Nadooshan et al. [16] have reported a coplanar 2:1 QCA MUX circuit. This circuit requires 26 cells and 0.02 μm^2 area. Sen et al. [17] have offered a multilayer layer 2:1 QCA MUX circuit. This circuit requires 23 cells and 0.01 μm^2 area. Sen et al. [17] have also proposed a multilayer layer 2:1 QCA MUX circuit. This circuit has 22 cells and 0.01 μm^2 area. Askari et al. [18] have reported a single layer 2:1 QCA MUX circuit. This circuit requires 35 cells and 0.04 μm^2 area. Asfestani et al. [19] have reported a coplanar 2:1 QCA MUX circuit. This circuit has 12 cells and 0.01 μm^2 area. Teodosio et al. [22] have reported two multilayer 2:1 QCA MUX circuits. These circuits consist of 0.14 μm^2(88) and 0.28μm^2 (146) area (cells). Rashidi et al. [23] have also reported another coplanar 2:1 QCA MUX circuit that has 16 cells and 0.01 μm^2 area. Kim et al. [30] have reported a coplanar 2:1 QCA MUX circuit. This circuit requires 46 cells and 0.08 μm^2 area.

These QCA MUX circuits provides suitable efficiency, but the efficiency of the QCA MUX circuit can be improved as described in the next sections.

5.3   THE DEVELOPED MUX CIRCUITS

This chapter presents novel 2^n:1 QCA MUX circuit for n = 1, 2, and 3.

5.3.1   THE DESIGNED 2^n:1 QCA MUX CIRCUIT FOR n = 1

Figure 5.8 displays the designed 2^n:1 QCA MUX circuit for n = 1.

This circuit is composed of an OR gate, an inverter gate, and two AND gates. In this circuit, the output will display the input A, when the address line, S, is equal to 0 and it will display the input B, when S = 1. This circuit has 0.01 μm^2 area and 15 cells.

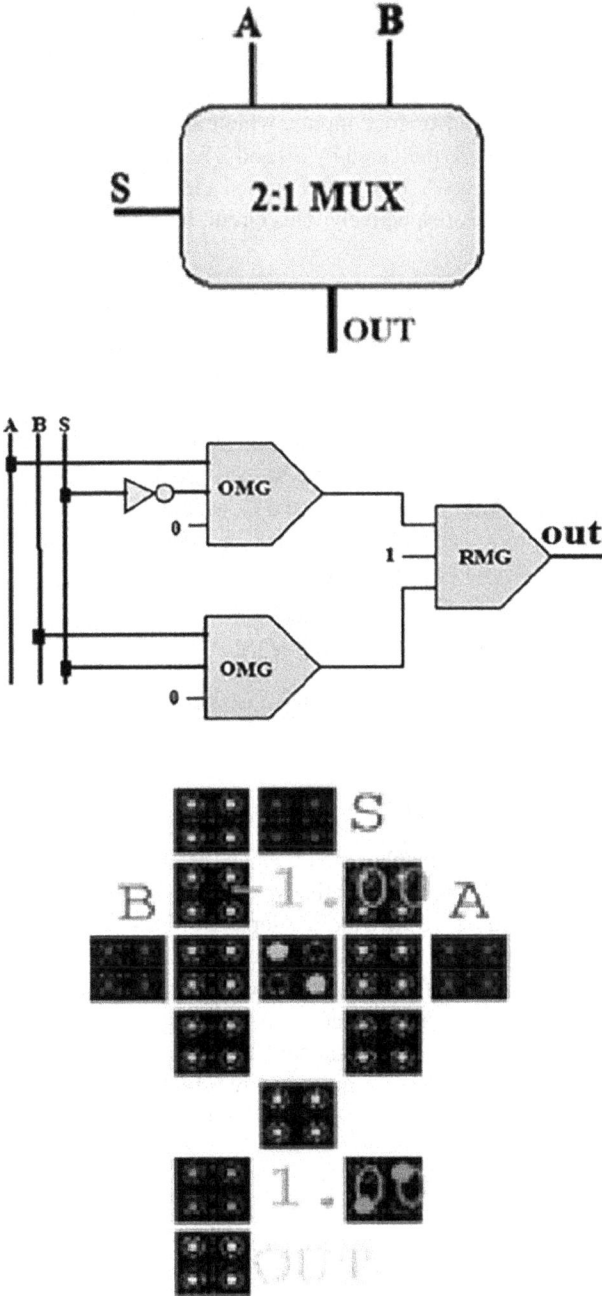

FIGURE 5.8 The designed 2^n:1 MUX circuit for n = 1 (a) Diagram (b) Circuit (c) QCA implementation.

5.3.2   The Designed 2^n:1 QCA MUX Circuit for n = 2

Figure 5.9 displays the designed 2^n:1 QCA MUX circuit for n = 2. This QCA MUX circuit is composed of three designed 2:1 QCA MUX circuit in this chapter.

This circuit is composed of four inputs, which are indicated by A, B, C, and D, two address lines, which are indicated by S0 and S1, and an output. In this circuit, the output will display the inputs A, B, C, and D, when address lines S0S1 = 00, S0S1 = 10, S0S1 = 11, and S0S1 = 01, respectively. This circuit has 0.05 μm² area and 50 cells.

5.3.3   The Designed 2^n:1 QCA MUX Circuit for n = 3

Figure 5.10 displays the designed 2^n:1 QCA MUX circuit for n = 3. This QCA MUX circuit is composed of the two developed 4:1 QCA MUX circuit and a 2:1 MUX circuit, which are developed in this chapter.

This circuit has eight inputs, which are indicated by A, B, C, D, E, F, G, and H, three address lines, which are indicated by S0, S1, and S2, and an output. The output will display the inputs A, B, C, D, E, F, G, and H, when S2S1S0 = 000, S2S1S0 = 001, S2S1S0 = 010, S2S1S0 = 011, S2S1S0 = 100, S2S1S0 = 101, S2S1S0 = 111, and S2S1S0 = 101.This circuit requires 0.18 μm² area and 159 cells.

5.4   THE RESULTS AND COMPARISON

The performance of the designed 2^n:1 MUX circuits is evaluated for n = 1, 2, and 3 using QCADesigner tool version 2.0.3.

5.4.1   The Simulation Results

Figures 5.11, 5.12, and 5.13 display the obtained results of the developed 2^n:1 QCA MUX circuit for n = 1, 2, and 3, respectively.

Based on the obtained waveforms, the developed 2^n:1 QCA MUX circuits for n = 1, 2, and 3 have ignorable distortion. In addition, these results confirm the functionality of our 2^n:1 QCA multiplexer circuits for n = 1, 2, and 3. For example, based on the waveforms in Figure 5.11, when address line S is equal to 0, the input A appears in output that is indicated by OUT. The input B appears in output OUT, when the address line S is equal to 1. In addition, based on the waveforms in Figure 5.12, the output OUT is shown input A, B, C, and D when the address lines S1S0 are 00, 10, 01, and 11, respectively.

5.4.2   The Comparison

Tables 5.1, 5.2 and 5.3 summarize the comparison between the developed 2^n:1 QCA MUX circuits compared to other QCA MUX circuits, for n = 1, 2, and 3, respectively. The costs in this chapter are calculated as follows.

$$Cos\,t1 = Area \times Latency \qquad (3)$$

$$Cost2 = Area \times Latency^2 \qquad (4)$$

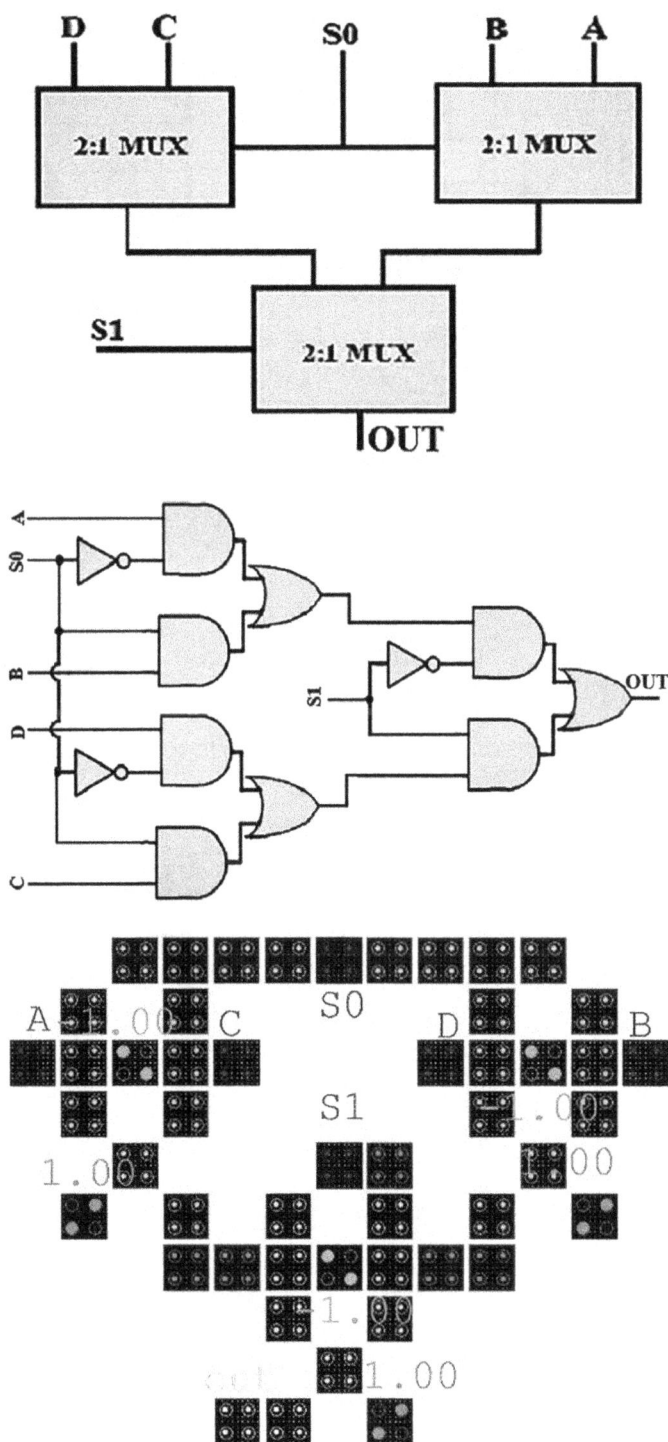

FIGURE 5.9   The designed $2^n{:}1$ QCA MUX circuit for n = 2 (a) Diagram (b) Circuit (c) QCA implementation.

S2	S1	S0	Out
0	0	0	A
0	0	1	B
0	1	0	C
0	1	1	D
1	0	0	E
1	0	1	F
1	1	0	H
1	1	1	G

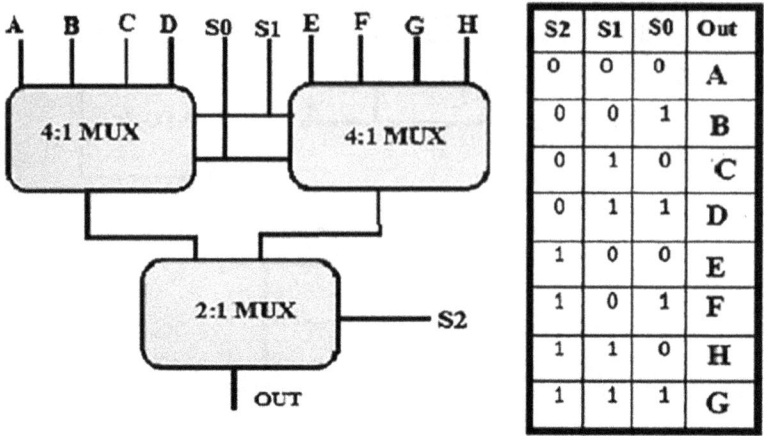

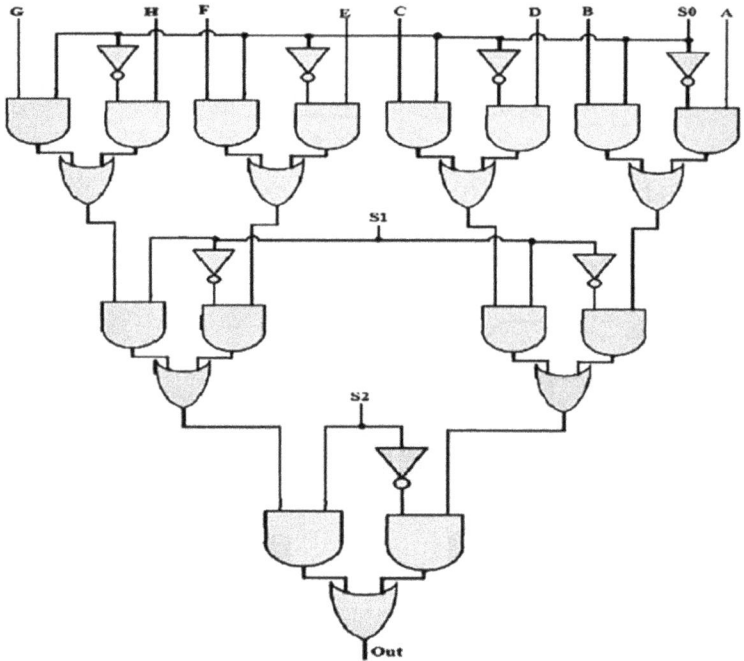

FIGURE 5.10 The Designed $2^n{:}1$ QCA MUX Circuit for n = 3 (a) Diagram (b) Circuit (c) QCA Implementation.

Where Area and Latency are shown based on μm^2 and clock phase.

Based on these results, the designed $2^n{:}1$ QCA MUX circuit provides benefits compared to $2^n{:}1$ QCA MUX circuits in [8–13, 15, 16, 18, 22, 30] for n = 1, [8, 10, 11, 14, 16–18, 20] for n = 2, and [11, 16, 20, 21] for n = 3 regarding complexity or cell count, area, latency, and costs.

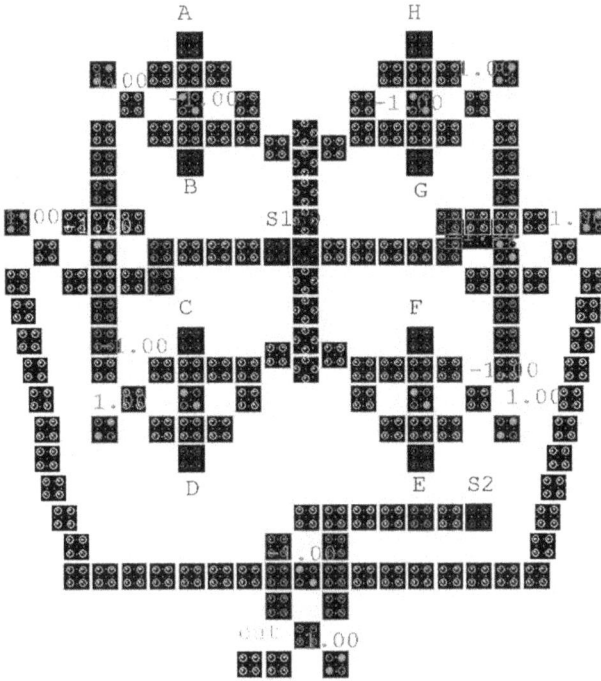

FIGURE 5.10 (Continued)

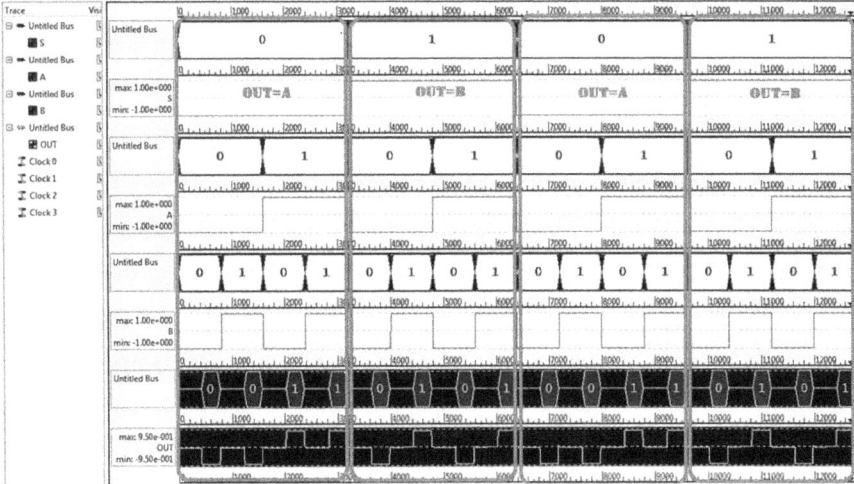

FIGURE 5.11 The Simulation Waveforms of Our 2^n:1 QCA Multiplexer Circuit for n = 1.

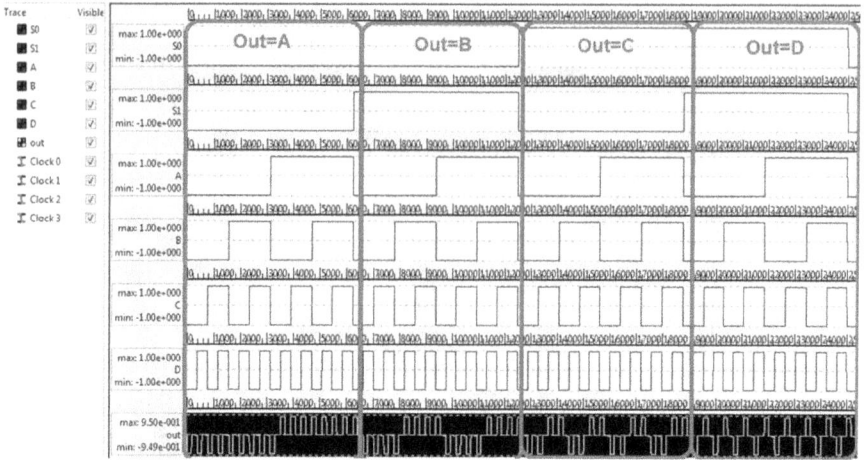

FIGURE 5.12   The Simulation Waveforms of Our $2^n{:}1$ QCA Multiplexer Circuit for n = 2.

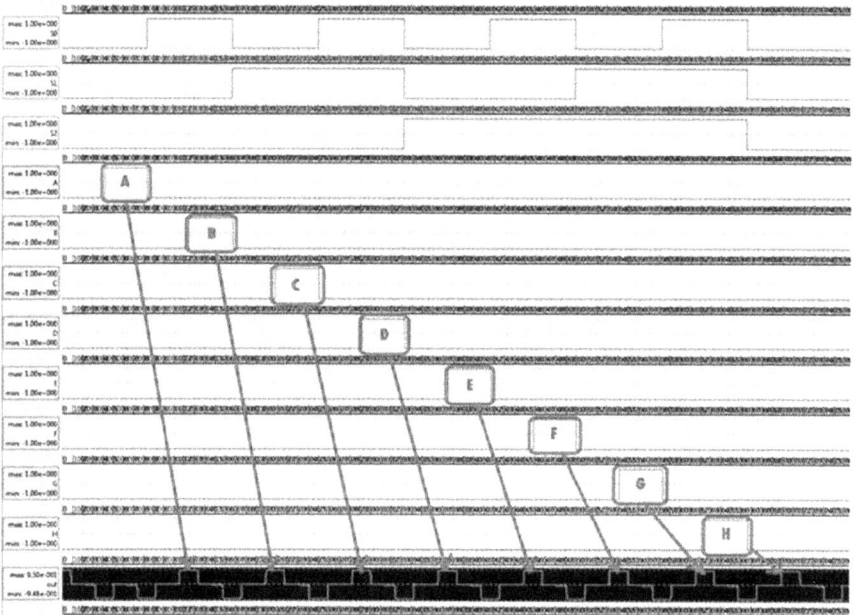

FIGURE 5.13   The Simulation Waveforms of Our $2^n{:}1$ QCA Multiplexer Circuit for n = 3.

The 2:1 QCA MUX circuit provided in [7] has similar results to our work, but the design of our 2:1 QCA MUX circuit is such that it provides better results for implementing $2^n{:}1$ QCA MUX circuit. The only 2:1 QCA MUX circuit that has better results than our work is the 2:1 QCA MUX circuit provided in [19]. It should be noted

TABLE 5.1
The simulation results for 2^n:1 QCA MUX circuits for n = 1

Reference	Complexity	Ratio	Area (μm^2)	Ratio	Latency (clock phase)	Ratio	Cost1	Ratio	Cost2	Ratio	Cross wiring
[22]	146	9.73	0.28	28	8	4	2.24	112	17.92	448	Multilayer
[22]	88	5.87	0.14	14	4	2	0.56	28	2.24	56	Multilayer
[30]	46	3.07	0.08	8	4	2	0.32	16	1.28	32	Coplanar
[10]	67	4.47	0.14	14	4	2	0.56	28	2.24	56	Coplanar
[9]	36	2.40	0.06	6	4	2	0.24	12	0.96	24	Multilayer
[18]	35	2.33	0.04	4	4	2	0.16	8	0.64	16	Coplanar
[11]	56	3.73	0.07	7	4	2	0.28	14	1.12	28	Coplanar
[12]	27	1.80	0.03	3	3	1.5	0.09	4.5	0.27	6.75	Coplanar
[13]	19	1.27	0.02	2	3	1.5	0.06	3	0.18	4.5	Coplanar
[16]	26	1.73	0.02	2	2	1	0.04	2	0.08	2	Coplanar
[15]	19	1.27	0.02	2	2	1	0.04	2	0.08	2	Coplanar
[17]	23	1.53	0.01	1	2	1	0.02	1	0.04	1	Multilayer
[17]	22	1.47	0.01	1	2	1	0.02	1	0.04	1	Multilayer
[8]	23	1.53	0.02	2	2	1	0.04	2	0.08	2	Coplanar
[7]	15	1.00	0.01	1	2	1	0.02	1	0.04	1	Coplanar
[19]	12	0.80	0.01	1	1	0.5	0.01	0.5	0.01	0.25	Coplanar
[23]	16	1.07	0.01	1	2	1	0.02	1	0.04	1	Coplanar
This chapter	15	1.00	0.01	1	2	1	0.02	1	0.04	1	Coplanar

TABLE 5.2
The simulation results for 2^n:1 QCA MUX circuits for n = 2

Reference	Complexity	Ratio	Area (μm^2)	Ratio	Latency (clock phase)	Ratio	Cost1	Ratio	Cost2	Ratio	Cross wiring
[10]	215	4.3	0.25	5	6	1.5	1.5	7.5	9	11.25	Coplanar
[20]	223	4.46	0.22	4.4	6	1.5	1.32	6.6	7.92	9.9	Multilayer
[18]	124	2.48	0.25	5	8	2	2	10	16	20	Coplanar
[11]	290	5.8	0.35	7	7	1.75	2.45	12.25	17.15	21.44	Coplanar
[16]	271	5.42	0.37	7.4	19	4.75	7.03	35.15	133.6	166.96	Coplanar
[17]	103	2.06	0.08	1.6	7	1.75	0.56	2.8	3.92	4.9	Multilayer
[17]	94	1.88	0.07	1.4	6	1.5	0.42	2.1	2.52	3.15	Multilayer
[8]	155	3.1	0.24	4.8	5	1.25	1.2	6	6	7.5	Coplanar
[14]	251	5.02	0.2	4	5	1.25	1	5	5	6.25	Multilayer
[14]	199	3.98	0.27	5.4	6	1.5	1.62	8.1	9.72	12.15	Coplanar
[7]	107	2.14	0.15	3	4	1	0.6	3	2.4	3	Coplanar
[19]	61	1.22	0.08	1.6	4	1	0.32	1.6	1.28	1.6	Coplanar
[23]	96	1.92	0.11	2.2	4	1	0.44	2.2	1.76	2.2	Coplanar
This chapter	50	1	0.05	1	4	1	0.2	1	0.8	1	Coplanar

TABLE 5.3
The simulation results for 2^n:1 QCA MUX circuits for n = 3

Reference	Complexity	Ratio	Area (μm^2)	Ratio	Latency (clock phase)	Ratio	Cost1	Ratio	Cost2	Ratio	Cross wiring
[20]	576	3.62	0.82	4.56	9	1.29	7.38	5.86	66.42	7.53	Coplanar
[11]	633	3.98	0.67	3.72	11	1.57	7.37	5.85	81.07	9.19	Coplanar
[16]	1312	8.25	1.83	10.17	42	6.00	76.86	61.00	3228	365.99	Coplanar
[8]	462	2.91	0.87	4.83	7	1.00	6.09	4.83	42.63	4.83	Coplanar
[21]	608	3.82	0.71	3.94	9	1.29	6.39	5.07	57.51	6.52	Coplanar
[21]	494	3.11	0.58	3.22	9	1.29	5.22	4.14	46.98	5.33	Coplanar
[7]	293	1.84	0.58	3.22	6	0.86	3.48	2.76	20.88	2.37	Coplanar
[19]	175	1.10	0.24	1.33	6	0.86	1.44	1.14	8.64	0.98	Coplanar
[23]	286	1.80	0.43	2.39	6	0.86	2.58	2.05	15.48	1.76	Coplanar
This chapter	159	1	0.18	1	7	1	1.26	1	8.82	1	Coplanar

that the design of our MUX circuit is such that it provides better results for 2^n:1 MUX circuit implementation.

The only 4:1 QCA MUX circuits that have similar latency with our 4:1 QCA MUX circuits are the 4:1 QCA MUX circuits provided in [7], [19] and [23]. It should be noted that the cell count, area, cost1, and cost2 in [7] are about 2, 2.5, 2.5 and 2.5 times, respectively, compared to our 4:1 QCA MUX circuit. The cell count, area, cost1, and cost2 in [19] are about 1.15, 1.33, 1.33 and 1.33 times, respectively, superior than our 4:1 QCA MUX circuit. In addition, the cell count, area, and cost1, and cost2 in [23] are about 1.81, 1.83, 1.83 and 1.83 times, respectively, compared to the proposed 4:1 QCA MUX circuit.

The only 8:1 QCA MUX circuit that has similar latency with our 8:1 QCA MUX circuit is the 8:1 QCA MUX circuits provided in [8]. It should be noted that the area, cost1, cost2, and cell count in [8] are by about 4.83, 4.83, 4.83 and 2.83 times, respectively, compared to our 8:1 QCA MUX circuit. Moreover, the only 8:1 QCA MUX circuits that have better latency compared to our proposed 4:1 QCA MUX circuit are the 8:1 QCA MUX circuits provided in [7], [19] and [23]. It should be noted that the area, cell count, cost1 and cost2 in [7] are about 3.22, 1.79, 2.76, and 2.37 times, respectively, compared to the designed 8:1 QCA MUX circuit. The cell count, area, and cost1 in [19] are about 1.07, 1.33 and 1.14 times, respectively, compared to our 8:1 QCA MUX circuit. Moreover, the cell count, area, cost1 and cost2 in [23] are about 1.75, 2.39, 2.06 and 1.76 times, respectively, compared to our 8:1 QCA MUX circuit.

5.5 CONCLUSIONS

The QCA technology is a novel nanotechnology that is utilized in digital circuits design. This technology is one kind of nanotechnology that may be used to replace conventional CMOS technology. On the other hand, the multiplexer circuit is a basic component in circuits design. This chapter presented a new and efficient 2:1 QCA MUX circuit that is composed of an OR gate, an inverter gate, and two AND gates. This QCA circuit can be extended to construct 2^n:1 QCA MUX circuit. The proposed 2^n:1 QCA MUX circuit for n = 2 was developed based on three our 2:1 QCA MUX, and the developed 2^n:1 QCA MUX circuit for n = 3 was developed based on two developed 4:1 and a developed 2:1 QCA MUX circuits. The design strategy in the proposed 2^n:1 QCA MUX circuits was using a cost-effective architecture and path-planning design, which can decrease the design costs. The QCADesigner tool version 2.0.3 was utilized for simulation of the designed QCA MUX circuits. The results demonstrated that the designed 2^n:1 QCA MUX circuits have 15 (0.01 μm^2), 50 (0.05 μm^2), and 159 (0.18 μm^2) cells (area) for n = 1, 2 and 3, respectively. The comparison results indicate that the designed 2^n:1 QCA MUX circuit provides benefits compared to 2^n:1 QCA MUX circuits in [8–13, 15, 16, 18, 22, 30] for n = 1, [8, 10, 11, 14, 16–18, 20] for n = 2, and [11, 16, 20, 21] for n = 3 regarding cell count, area, latency, and costs.

REFERENCES

1. Niknezhad Divshali, M., A. Rezai, and A. Karimi, *Novel multilayer SISO shift register architecture in QCA technology and its usage in communications.* International Journal of Communication Systems, 2022. **35**(8): p. e5121.

2. Ansarian Najafabadi, S., A. Rezai, and K. Ghasvarian Jahromi, *Novel circuit design for reversible multilayer ALU in QCA technology.* Journal of Computational Electronics, 2022. **21**: p. 1–10.

3. Sridharan, K. and V. Pudi, *Design of arithmetic circuits in quantum dot cellular automata nanotechnology.* Vol. 599. 2015: Springer.

4. Balali, M., et al., *Towards coplanar quantum-dot cellular automata adders based on efficient three-input XOR gate.* Results in Physics, 2017. **7**: pp. 1389–1395.

5. Shiri, A., A. Rezai, and H. Mahmoodian, *Design of efficient coplanar comprator circuit in QCA technology.* Facta Universitatis. Series: Electronics and Energetics, 2019. **32**(1): pp. 119–128.

6. Rashidi, H. and A. Rezai, *Design of novel efficient multiplexer architecture for quantum-dot cellular automata.* Journal of Nano- and Electronic Physics, 2017. **9**: 01012-1–01012-7.

7. Rashidi, H., A. Rezai, and S. Soltany, *High-performance multiplexer architecture for quantum-dot cellular automata.* Journal of Computational Electronics, 2016. **15**(3): pp. 968–981.

8. Sen, B., et al., *Towards modular design of reliable quantum-dot cellular automata logic circuit using multiplexers.* Computers and Electrical Engineering, 2015. **45**: pp. 42–54.

9. Hashemi, S., M.R. Azghadi, and A. Zakerolhosseini. *A novel QCA multiplexer design,* in *International Symposium on Telecommunications.* 2008. IEEE.

10. Mardiris, V., et al. *Design and simulation of a QCA 2 to 1 multiplexer,* in *International Conference on Computers.* 2008. Heraklion, Greece.

11. Mardiris, V.A. and I.G. Karafyllidis, *Design and simulation of modular 2n to 1 quantum-dot cellular automata (QCA) multiplexers.* International Journal of Circuit Theory and Applications, 2010. **38**(8): pp. 771–785.

12. Roohi, A., et al., *A novel architecture for quantum-dot cellular automata multiplexer.* International Journal of Computer Science Issues, 2011. **8**(1): pp. 55–60.
13. Sen, B., et al., *An efficient multiplexer in quantum-dot cellular automata*, in *Progress in VLSI Design and Test.* 2012, Springer. p. 350–351.
14. Cocorullo, G., et al., *Design of efficient QCA multiplexers.* International Journal of Circuit Theory and Applications, 2016. **44**(3): pp. 602–615.
15. Sen, B., et al., *Modular design of testable reversible ALU by QCA multiplexer with increase in programmability.* Microelectronics Journal, 2014. **45**(11): pp. 1522–1532.
16. Nadooshan, R.S. and M. Kianpour, *A novel QCA implementation of MUX-based universal shift register.* Journal of Computational Electronics, 2014. **13**(1): pp. 198–210.
17. Sen, B., et al., *Towards the hierarchical design of multilayer QCA logic circuit.* Journal of Computational Science, 2015. **11**: pp. 233–244.
18. Askari, M., M. Taghizadeh, and K. Fardad. *Digital design using quantum-dot cellular automata (a nanotechnology method)*, in *Proceedings of the International Conference on Computer and Communication Engineering.* 2008. Kuala Lumpur, Malaysia: IEEE.
19. Asfestani, M.N. and S.R. Heikalabad, *A unique structure for the multiplexer in quantum-dot cellular automata to create a revolution in design of nanostructures.* Physica B, 2017. **512**: pp. 91–99.
20. Vankamamidi, V., M. Ottavi, and F. Lombardi, *Two-dimensional schemes for clocking/timing of QCA circuits.* IEEE Transactions on Computer-Aided Design of Integrated Circuits and Systems, 2007. **27**(1): pp. 34–44.
21. Tung, C.-C., R.B. Rungta, and E.R. Peskin. *Simulation of a QCA-based CLB and a multi-CLB application*, in *International Conference on Field-Programmable Technology.* 2009. United States of America: IEEE.
22. Teodósio, T. and L. Sousa. *QCA-LG: A tool for the automatic layout generation of QCA combinational circuits*, in *Norchip.* 2007. IEEE.
23. Rashidi, H. and A. Rezai, *Design of novel multiplexer circuits in QCA nanocomputing.* Facta Universitatis, Series: Electronics and Energetics, 2021. **34**(1): pp. 105–114.
24. Sen, B., A. Rajoria, and B.K. Sikdar, *Design of efficient full adder in quantum-dot cellular automata.* The Scientific World Journal, 2013. **2013**: p. 250802.
25. Roshany, H.R. and A. Rezai, *Novel efficient circuit design for multilayer QCA RCA.* International Journal of Theoretical Physics, 2019. **58**(6): pp. 1745–1757.
26. Kim, K., K. Wu, and R. Karri. *Towards designing robust QCA architectures in the presence of sneak noise paths*, in *Proceedings of the conference on Design, Automation and Test in Europe.* 2005. IEEE Computer Society.
27. Adelnia, Y. and A. Rezai, *A novel adder circuit design in quantum-dot cellular automata technology.* International Journal of Theoretical Physics, 2019. **58**(1): pp. 184–200.
28. Enayati, M., A. Rezai, and A. Karimi, *Efficient circuit design for content-addressable memory in quantum-dot cellular automata technology.* SN Applied Sciences, 2021. **3**(10): pp. 1–10.
29. Divshali, M.N. and A. Rezai, *Novel circuits design for SISO shift register in QCA technology.* Journal of Circuits, Systems and Computers, 2021. **30**(11): p. 2150203.
30. Kim, K., et al., *The robust QCA adder designs using composable QCA building blocks.* IEEE Transactions on Computer-Aided Design of Integrated Circuits and Systems, 2006. **26**(1): pp. 176–183.

6 An Optimized Approach of Designing Register and Counter in QCA

Vaibhav Jain[a], Devendra Kumar Sharma[b] and Hari Mohan Gaur[c]

[a]Department of Electronics and Communication Engineering, ABES Institute of Technology, Ghaziabad, India; [a,b]Department of Electronics and Communication Engineering, SRM Institute of Science and Technology, NCR Campus, Ghaziabad, India;
[c]School of Computer Science Engineering and Technology, Bennett University, Greater Noida, India.

6.1 INTRODUCTION

The CMOS technology has now reached to its final limits in the development of reduce size and low speed circuit. The various responsible factors include large power dissipation, trade off between size reduction, and low cost with failure of Denard Scaling [Mack (2015)]. In past decades, several efforts have been made from the researchers to bring out an alternative solution which replaces CMOS technology. Quantum-Dot Cellular Automata (QCA) is found to be the best replacement of the CMOS technology [Gaur et al. (2020) Gaur, Sasamal, Singh, Mohan, and Pradhan]. It was invented in 1993 by Lent et al. in the form of new computing paradigms, operates at nanolevel, and physically implemented first time in 1997 [Tougaw and Lent (1994)]. In QCA, the signal propagates according to Coulomb's law as there is no sign of current flow. The technology revolves around its fundamental block called quantum cell which is of small size as compared to traditional transistors. Therefore, the circuits designed using this technology are smaller in size and higher in speed. In addition, reversible gates can also be considered as an alternative solution to reduce energy dissipation [Gaur and Singh (2016), Gaur et al. (2015) Gaur, Singh, and Ghanekar].

In digital electronics, circuit designing requires AND, OR, NOT logic gates at primary level. To perform these operations, a number of quantum cells are arranged to form three primary primitives design commonly known as 3-input majority gate, wire, and inverter. In addition, a clock signal is essential to control and synchronize the operation of the quantum circuit. These primitives along with clock signal defines

DOI: 10.1201/9781003361633-6

complexity of the designed circuit. Therefore, researchers focus continually to reduce the utilization of majority gate, inverters, delay, and crossovers. Cell interaction approach minimizes the complexity at large level, hence widely used in circuit designing. Here, cells are placed in random manner which polarized its neighbor cell to generate the required output. The importance of combinational along with sequential logic circuits in designing of digital processor cannot be neglected. This chapter represents few sequential circuit designs in QCA technology such as D flip-flop, T flip-flop, Register, and Counter of length 2-bit. The proposed D flip-flop takes 27 cells in an area of 0.03 μm^2 and T flip-flop uses 20 cells in 0.02 μm^2 area. The 2-bit Parallel-In-Parallel-Out (PIPO) register design build with 56 cells and occupies 0.05 μm^2 area whereas utilizing 47 cells in similar area 2-bit counter circuit have been presented.

In literature, several authors have been reported D and T flip-flop architectures using different combinations of 3-input majority gate and inverters [Vetteth et al. (2003) Vetteth, Walus, Dimitrov, and Jullien; Hashemi and Navi (2012); Goswami et al. (2014) Goswami, Kumar, Tibrewal, and Mazumdar; Abutaleb (2017); Sasamal et al. (2019) Sasamal, Singh, and Ghanekar; Alamdar et al. (2020) Alamdar, Ardeshir, and Gholami; Alamdar et al. (2021) Alamdar, Ardeshir, and Gholami; Purkayastha et al. (2018) Purkayastha, De, and Chattopadhyay; Nafees et al. (2022) Nafees, Ahmed, Kakkar, Bahar, Wahid, and Otsuki; Chakrabarty et al. (2018) Chakrabarty, Mahato, Banerjee, Choudhuri, Dey, and Mandal; Torabi(2011); Angizi et al. (2014) Angizi, Navi, Sayedsalehi, and Navin; Angizi et al. (2015) Angizi, Moaiyeri, Farrokhi, Navi, and Bagherzadeh]. Using twelve 3-input majority gate with four inverters, the 4-bit PIPO shift register designs were also reported [Purkayastha et al. (2018) Purkayastha, De, and Chattopadhyay; Nafees et al. (2022) Nafees, Ahmed, Kakkar, Bahar, Wahid, and Otsuki; Jeon(2020)]. During the last decade, the reported D flip-flop designs utilize a maximum of 74 cells whereas 21 is the minimum. The latency is found to be varying from a value of 1.5 to 0.5 using single layer crossover. However, the space requirement by the designs starts with 0.10 μm^2 and ends with 0.02 μm^2. It can be seen that complex T flip-flop designs were reported with 92 cells in an area of 0.10 μm^2 which takes 05 clock zones to generate the required output. Further reduction leads to the development of a less complex T flip-flop circuit utilizing only one 3-input majority gate. It consists of 21 cells in 0.02 μm^2 area and produces output after 02 clock zones only. The reduced design complexity also decrement cost-function value to 0.5. The development of 4-bit PIPO circuits starts with 260 cells in an area of 1.67 μm^2 and requires 04 clock zones for output generation. The optimized design has similar speed but reduces the utilization of cell count to value 114. It is also found in literature that several authors reported counter designs in varying length which utilize T flip-flop at primary level. [Angizi et al. (2015) Angizi, Moaiyeri, Farrokhi, Navi, and Bagherzadeh; Majeed et al. (2019) Majeed, Alkaldy, bin Zainal, Nor, et al.; Yang et al. (2010) Yang, Cai, Zhao, and Zhang; Sheikhfaal et al. (2015) Sheikhfaal, Navi, Angizi, and Navin; Bhavani and Alinvinisha (2015); Sangsefidi et al. (2018) Sangsefidi, Abedi, Yoosefi, and Karimpour]. The development starts with utilization of maximum fourteen 3-input majority gate with six inverters and two crossover. It is built using 328 quantum cells in an area of 0.62 μm^2 with a delay of three clock cycles. The optimized higher speed design which produces output after 02 clock zones only occupies 78 cells in an area of 0.06 μm^2.

The chapter organization is as follows: Section 2 gives information essential to understanding QCA technology. The proposed flip-flop, register, and counter design based on cell interaction approach is discussed in Section 3. The simulation results along with comparison with previous work is also done in Section 3. Section 4 demonstrates energy estimation for each proposed design and the conclusions have been presented in Section 5.

6.2   QCA TERMINOLOGY

To start with QCA, it's essential to understand the basic terms utilized in designing and implementation. This section begins with the introduction to quantum cell that propagate information from source to receiver. The other primitives called majority gate, inverter, and wire along with design parameters and power dissipation are also elaborated.

(1) **QCA Cell:** It's a square shape with four dots placed in the respective corners and populated with two electrons as shown in Figure 6.1. As per Coulomb's law, the electrons will always settle in one of the 02 possible diagonal positions inside a cell. The electron movement is dependent upon the energy barrier level between the adjacent dots. Hence, a cell always remains in any one of the possible polarization states +1 and −1 which represent logic state 1 and 0 respectively. These states are illustrated in Figure 6.1.

(2) **QCA Wire:** A horizontal or vertical line of cells placed one after another represents QCA wire as shown in Figure 6.2. It is used to transfer same digital information from input to output cell. For instance, here the input cell is polarized as logic 1 and the same is transferred to the output cell.

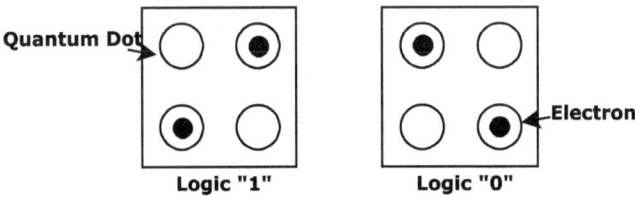

FIGURE 6.1   QCA cell.

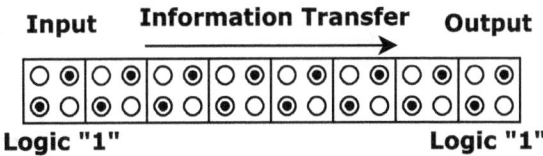

FIGURE 6.2   QCA wire.

(3) **QCA Inverter:** The complement of digital input is required at various stages in designing. To achieve this cells initially are placed as illustrated in Figure 6.3. Presently inversion can be performed by placing a cell diagonal to its neighbor cell as shown in Figure 6.4.

(4) **3-input Majority Gate:** Other than NOT gate, the next essential requirement is the designing of basic gates (AND/OR) that perform addition and subtraction operations. To accomplish the task, two layouts called 3-input majority gate (MV3) and rotating majority gate (RMV3) have been designed by placing the cells as illustrated in Figure 6.5 and Figure 6.6. Each layout contains three input cells, one output cell, and the remaining is called the device cell. The polarization of the output cell is set to logic HIGH when two or more inputs out of three is polarized with logic HIGH. Consider A, B, and C represents input cell variables then the output of MV3 and RMV3 is calculated as per Eq. 1 and Eq. 2 respectively.

$$3_{QCA} = AB + BC + AC \tag{1}$$

$$RMV\,3_{QCA} = AB + BC + AC \tag{2}$$

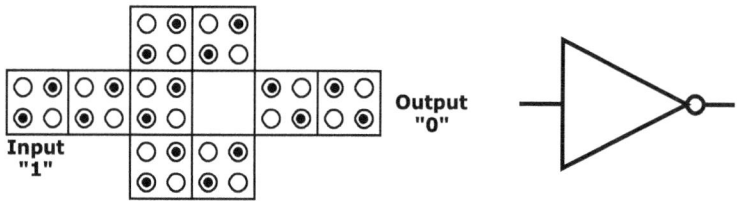

FIGURE 6.3   QCA inverter.

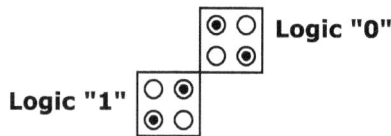

FIGURE 6.4   Optimized QCA inverter.

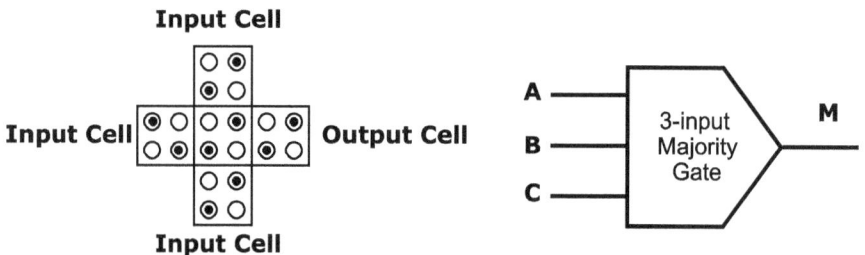

FIGURE 6.5   QCA 3-input majority gate.

Input Cell   **Input Cell**

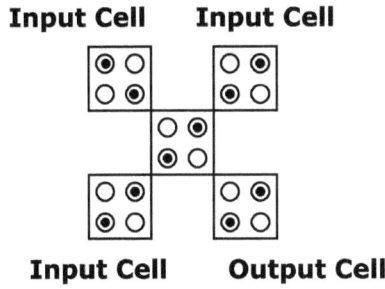

Input Cell   **Output Cell**

FIGURE 6.6   QCA 3-input rotating majority gate.

TABLE 6.1
Truth table of 3-input majority gate

Input			Output
A	B	C	OUT
0	0	0	0
0	0	1	0
0	1	0	0
0	1	1	1
1	0	0	0
1	0	1	1
1	1	0	1
1	1	1	1

Using any majority gate, AND operation can be performed by fixing one of the input cells to logic state 0. However to perform OR, one input cell is kept fixed at logic state 1 in both the layouts. The truth table illustrating all possible inputs and respective outputs with symbol has been demonstrated in Table 6.1 and Figure 6.5 respectively.

(5) **QCA Clocking:** The circuit layout built in QCA technology have not used external power supply. Hence, results in less power dissipation. Also, information transfer does not involve flow of current which gives rise to the need of a control signal for synchronization. To serve this need, a clock is required which also supplies real power for the operation of circuit [Tougaw and Lent (1994), Tougaw and Lent (1996)]. It can be found that the most popular clocking scheme used in literature, is adiabatic switching. In this, a clock cycle is further classified into four clock phases/zones termed as switch, hold, release, and relax respectively. The potential levels in each clock zone at different times is illustrated in Figure 6.7. Each clock phase differs from its adjacent zone by phase difference of $\pi/2$ [Lent and Tougaw (1997)].

(6) **Performance Parameters:** The parameters used to compare different designed circuit in QCA technology are termed as performance metric. Cell count, area, latency, and cost-function are found to be widely used metrics in

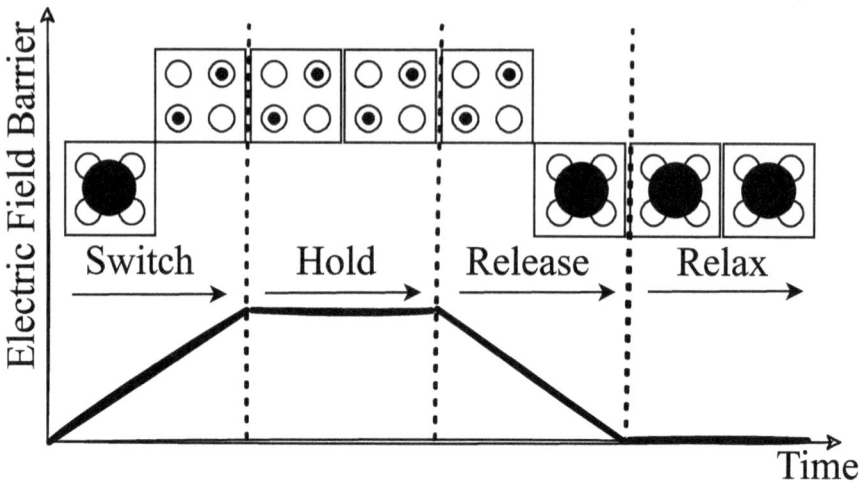

FIGURE 6.7   Energy barrier vs time diagram in QCA clocking.

literature. The total number of cells utilized in the design refer as cell count. Area is the total space covered by the cells used in the layout. For circuit delay computation, also referred to as latency, depends on the count of clock zones required to generate output and calculated by [(1/4) * total clock phase count]. The measurement of complexity is called cost function that depends on the count of majority gate, inverters, crossover, and latency [Liu et al. (2014) Liu, Lu, ONeill, and Swartz- lander]. It is calculated for the proposed and previous architectures using the formula defined in Eq. 3.

$$Cost_{QCA} = (M^k + I + C^l) * T^p \qquad (3)$$

where M indicates the number of majority gates, I are the inverters, C represents crossover, and T shows latency of the circuit. Also, k, l and p are coefficients. In general, the value of k and l is 2 and the value of p is 1.

(7) **Power Dissipation:** The changing technology era requires reduced area, delay, and optimized cost design. In addition, no external supply is used in QCA layout that results in low power dissipation. Hence, power estimation also becomes a popular metric used to measure the circuit efficiency. QCA Designer-E tool is widely used to compute total and average power per cycle in eV. The tool runs in coherence vector energy simulation environment and initialized with 500,000 samples out of which 3,000 will be recorded for measurement [Torres et al. (2018) Torres, Wille, Niemann, and Drechsler]. In addition, simulation results in iterations count required to converge the initial steady state polarization with simulation time.

Another tool called QCA Pro has been utilized for the measurement of polariza- tion error and non-adiabatic switching power loss at 0.5Ek, 1.0Ek and 1.5Ek tun- neling energy levels.

6.3 SEQUENTIAL LOGIC CIRCUITS IN QCA TECHNOLOGY

In literature, several designs of sequential circuits have been reported in QCA technology. Here, the output depends on the present input along with previous value of output. Flip-flop is a primary block utilized in the designing of any sequential circuit which is used to store 1-bit of information. The value stored in flip-flop will change by either level or edge triggered clock signal. It is also found that three types of approaches are utilized in QCA designs namely majority gate, reversible gate, and cell interaction. This section describes the design of D and T flip-flop at primary level. In addition, register, and counter circuit using D and T flip-flop is also proposed based on cell interaction approach. The presented architectures are level triggered and use the layout of previously reported Exclusive-OR gate design [Jain et al. (2022) Jain, Sharma, and Gaur].

6.3.1 D FLIP-FLOP

The level triggered D flip-flop (DFF) circuit is used to hold a bit of information. Here, the output holds its previous value if the clock or enable signal is at logic 0. Whereas, when clock signal changes its state to logic 1 the same input information appears on output. The logic diagram reflects its operation is shown in Figure 6.8 and respective QCA layout is illustrated in Figure 6.9. Consider D as input, CLK as clock and $Q(t-1)$ as previous value of output, than Boolean function that defines the present output of flip-flop is listed in Eq. 4.

$$Q(t)_D = CLK * D + CLK^0 * Q(t-1) \tag{4}$$

The truth table that includes all the possible input combinations and their respective outputs has been listed in Table 6.2. Its proposed equivalent QCA layout consists of 27 cells in an area of 0.03 μm^2. The circuit utilizes only 03 clock zones to produce output, hence latency of the circuit is 0.75, also verified with the simulation results demonstrated in Figure 6.10. The designed DFF layout is also simulated on tool called QCA Designer-E for the measurement of power dissipation. It takes 08 iterations to converge the initial steady state polarization and 15 seconds as total

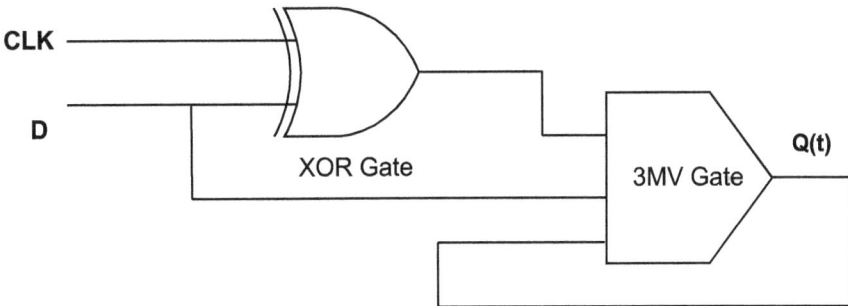

FIGURE 6.8   Logic diagram of D flip-flop.

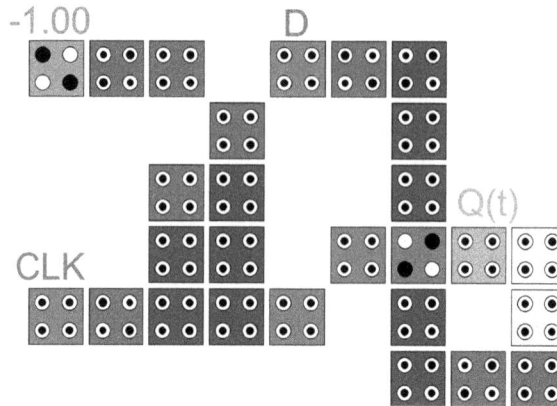

FIGURE 6.9   QCA layout of D flip-flop.

TABLE 6.2
Behavior of D flip-flop

Input (CLK)	Input (D)	Output (Q(t))
0	0	Q(t-1)
0	1	Q(t-1)
1	0	0
1	1	1

simulation time with 02 inputs and 01 output. The amount of total and average energy per cycle during the simulation has been tabulated in Table 6.3. It can be found in literature that several authors presented D flip-flop designs with variations of complexities and design parameters. These are summarized and compare against performance measures Area, cell count, delay and cost function as tabulated in Table 6.4.

It can be seen from Table 6.4 that the proposed structure proves itself superior in area with respect to DFF-1 to DFF-4, DFF-7, DFF-8, DFF-9, and DFF-11. The layout is designed with reduced cell count as compared to prior designs except DFF-6 and DFF-10. However, the proposed architecture is cost efficient as compared to designs DFF-1 to DFF-11. Utilizing three 3-input majority gate with one inverter, DFF-1 to DFF-5 have been reported. DFF-1 also uses one crossover and consists of 66 cells in an area of 0.08 μm^2. DFF-2 takes 48 cells which occupied 0.05 μm^2 area and produces output after 04 clock zones with cost-function 10. DFF-3 is highly optimized that uses only 30 cells and took 03 clock zones for output generation. DFF-4 have been represented with 74 cells in an area of 0.10 μm^2 in a 4 × 4 USE grid clocking scheme. The author also presents optimized DFF-5 which utilize only 28 cells in 0.02 μm^2 area. The design takes 02 clock zones to produce the required output and its cost-function results in value of 2.5. Utilizing 3-input majority gate with an inverter, DFF-6 have been reported. It consists of 23 cells, occupies 0.02 μm^2 area and

Simulation Results

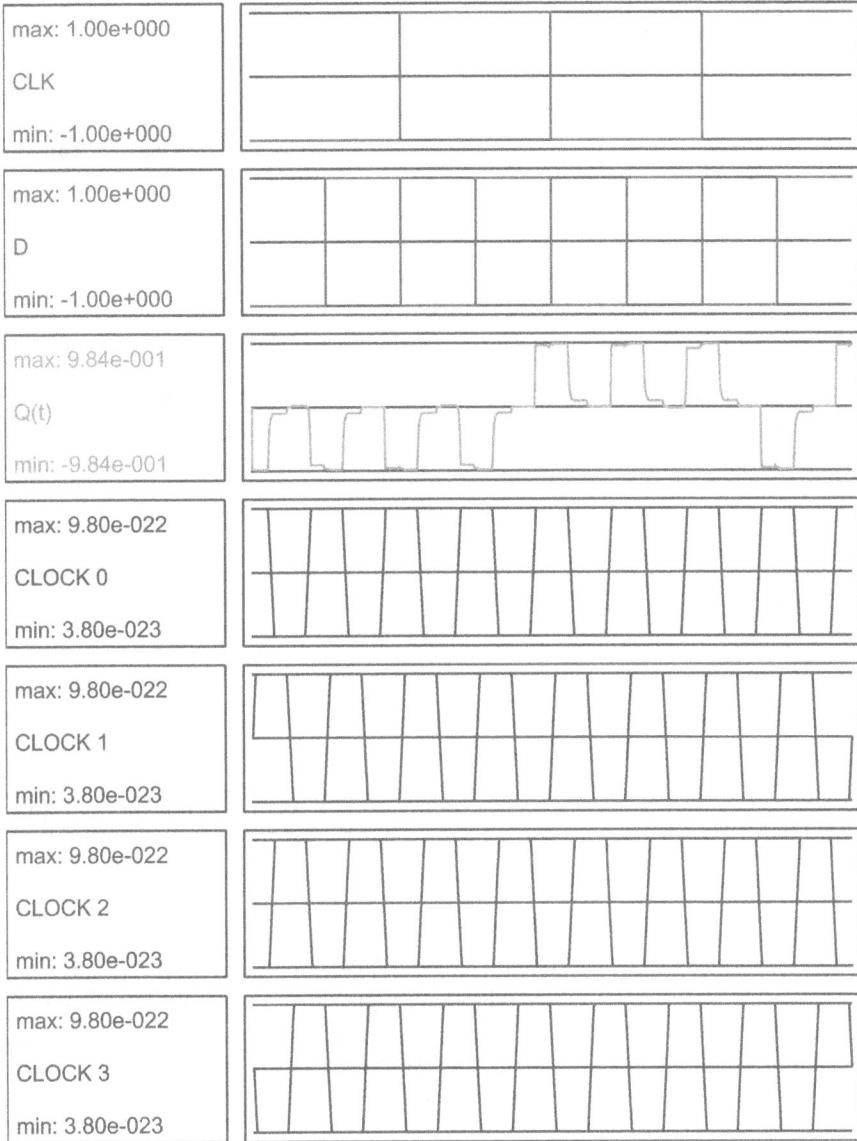

FIGURE 6.10   I/O waveform of D flip-flop.

TABLE 6.3
D flip-flop: power dissipation

Energy Category	Energy Dissipation Value in eV
Total energy dissipation (Sum Ebath)	1.57e-002
Average energy dissipation per cycle (Avg Ebath)	1.42e-003

TABLE 6.4
Comparison QCA D flip-flop

QCA DFF Designs	Area Occupied (μm^2)	No. of Cells	Delay (Clock Cycles)	Cost Function
DFF-1 [Vetteth et al. (2003) Vetteth, Walus, Dimitrov, and Jullien]	0.08	66	1.5	24.75
DFF-2 [Hashemi and Navi (2012)]	0.05	48	1	10.00
DFF-3 [Goswami et al. (2014) Goswami, Kumar, Tibrewal, and Mazumdar]	0.03	30	0.75	5.63
DFF-4 [Abutaleb (2017)]	0.10	74	1.5	22.50
DFF-5 [Abutaleb (2017)]	0.02	28	0.5	2.50
DFF-6 [Sasamal et al. (2019) Sasamal, Singh, and Ghanekar]	0.02	23	0.5	2.50
DFF-7 [Alamdar et al. (2020) Alamdar, Ardeshir, and Gholami]	0.08	59	1	10.00
DFF-8 [Alamdar et al. (2021) Alamdar, Ardeshir, and Gholami]	0.05	54	1.25	18.75
DFF-9 [Purkayastha et al. (2018) Purkayastha, De, and Chattopadhyay]	0.03	48	0.75	5.63
DFF-10 [Nafees et al. (2022) Nafees, Ahmed, Kakkar, Bahar, Wahid, and Otsuki]	0.02	21	1	10.00
DFF-11 [Chakrabarty et al. (2018) Chakrabarty, Mahato, Banerjee, Choudhuri, Dey, and Mandal]	0.33	43	1.25	26.56
Proposed DFF	0.03	27	0.75	1.13

generates output after 02 clock zones. DFF-7, DFF-8, and DFF-9 build up with 59, 54 and 48 cells in an area of 0.08 μm^2, 0.05 μm^2, 0.03 μm^2 respectively. As compared to DFF-7 and DFF-8, the design DFF-9 have higher speed as it takes 03 clock zones to produce output. Further DFF-10 have been reported with optimized cell count but high delay as it generates output after 04 clock zones. DFF-11 uses four 3-input majority gate and a inverter to produce output after 05 clock zones. It takes 43 cells which covers 0.033 μm^2 area.

6.3.2 T Flip-Flop

A T flip-flop (TFF) has only one input called T (toggle or trigger). It is also a single bit storage device and commonly used to generate the complement of past output value stored inside it. In the absence of clock signal, the output remains to its past value independent to input T. However, if clock signal is at logic "1", then output value changes to its complement for $T = 1$ and remains equal to its past state for $T = 0$.

Considering $Q(t-1)$ as previous output, T as input then the current value of output $Q(t)$ is generated using one 2-input EX-OR gate as shown in Figure 6.11. The output can be calculated using Boolean equation defined in Eq. 5.

$$Q(t)_T = T * Q(t-1)^t + T^t * Q(t-1)$$ (5)

The truth table showing all possible input and output combinations has been tabulated in Table 6.5. The proposed layout design in single layer which consists of 20 cells in an area of 0.02 μm^2 and produces output after 03 clock zones is represented in Figure 6.12. The simulation results that verify the operation of T flip-flop is illustrated in Figure 6.13. The power dissipation measurement is done using QCA Designer-E which takes 10 iterations to converge the initial steady state polarization. The total simulation time is found to be 11 seconds with single input and output. The total and average energy per cycle generated in simulation has been listed in Table 6.6. As per literature, several authors reported T flip-flop designs with variations of complexities and design parameters. The metrics are summarized and compared against previous work as illustrated in Table 6.7.

It can be seen from Table 6.7 that the proposed architecture shows its efficacy over prior structures in all performance metrics area, cell count, delay, and cost function. The design TFF-5 have been reported with higher speed and low cost but utilizes extra cells as compared to the proposed design. At the beginning, utilizing five 03-input majority gates with three inverters and one crossover TFF-1 was reported. The design complexity results in high cost-function 36.25, due to higher count of majority gate, inverters and crossover utilized. TFF-1 consists of 92 cells in 0.10 μm^2 area and uses 05 clock zones to produce output. Having similar speed TFF-2 have been reported that takes 66 cells, occupies 0.06 μm^2 area. The design presented

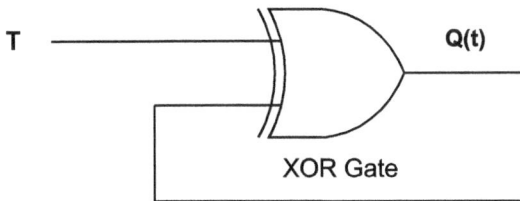

FIGURE 6.11   Logic diagram of T flip-flop.

TABLE 6.5
I/O behavior of T flip-flop

Previous Output [Q(t-1)]	Input (T)	Output (Q(t))
0	0	0
0	1	1
1	0	1
1	1	0

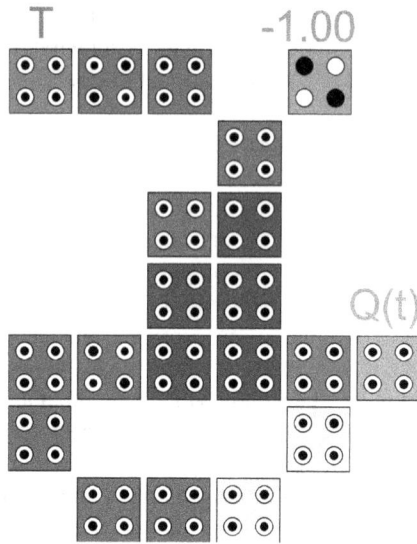

FIGURE 6.12   QCA layout of T flip-flop.

with reduced number of 3-input majority gate and inverters. TFF-3 and TFF-4 were reported with four 3-input majority gate and two inverters. TFF-3 consists of 55 cells and produce output after 06 clock zones. Whereas, TFF-4 takes only 46 cells and 04 clock zones to produce the required output. Further utilizing only 20 cells in 0.02 μm^2 area, TFF-5 was presented. The layout reported with less complexity as it uses cell interaction approach and only one majority gate. Consequently the cost function is also reduces to value 1.5.

6.3.3   REGISTER

The term register is used to store more than 1-bit of information. It is formed by placing multiple flip-flops adjacent to each other and transfer data in serial or parallel mode. Based upon configuration and data movement, registers are commonly called shift registers and are classified as Serial-In-Serial-Out (SISO), Serial-In-Parallel-Out (SIPO), Parallel-In-Serial-out (PISO), and Parallel-In-Parallel-Out (PIPO) shift register. D flip-flop is used at primary level in the designing of a shift register. Utilizing the design of DFF 3.1, 2-bit PIPO, and 4-bit PIPO shift registers have been proposed which uses two and four proposed DFF design respectively. A 2-bit PIPO shift register design at gate level using two DFF is illustrated in Figure 6.14. It can be seen that all the inputs are applied in parallel manner and the respective outputs are also generated at same time. Considering $D1$ and $D2$ as inputs, the Boolean expression of the required outputs is given in Eq. 6 and Eq. 7 respectively.

$$Q1 = CLK * D1 + CLK^t * Q1(t - 1) \tag{6}$$

$$Q2 = CLK * D2 + CLK^t * Q2(t - 1) \tag{7}$$

Simulation Results

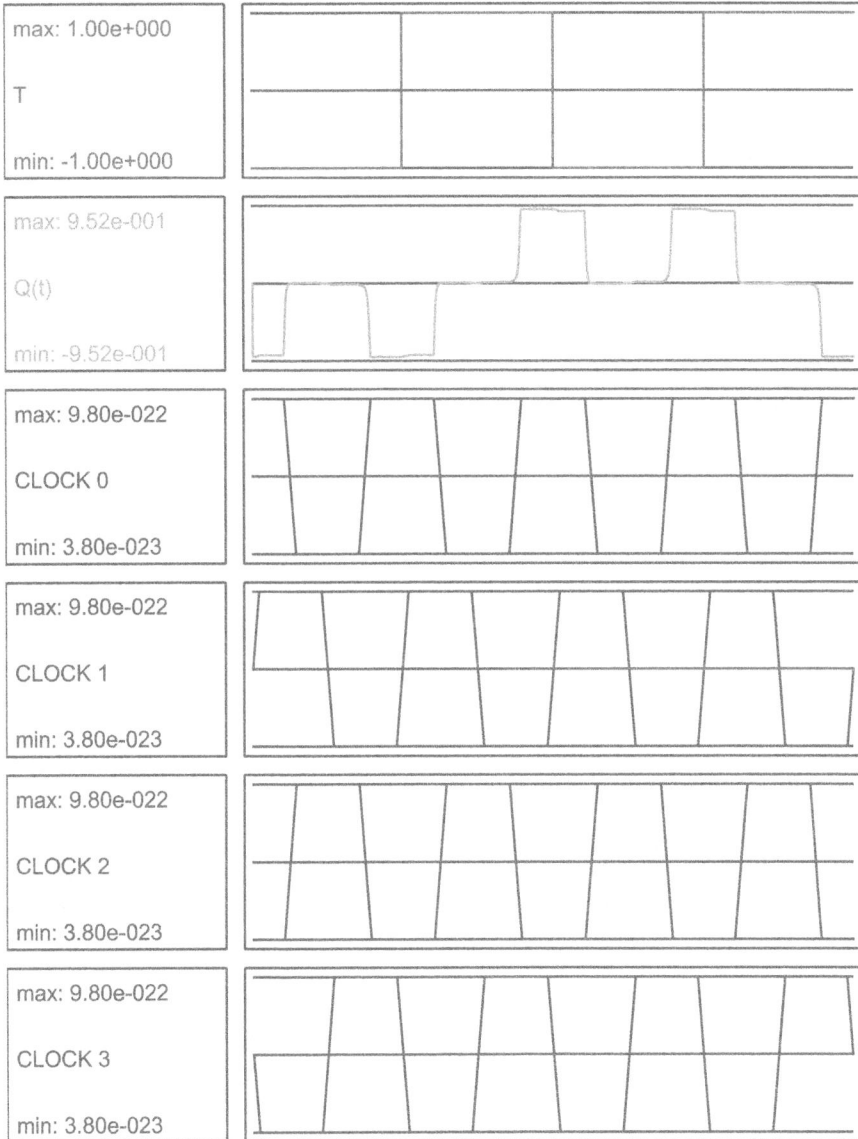

FIGURE 6.13   I/O Waveform of T flip-flop.

Here, $Q1(t - 1)$ and $Q2(t - 1)$ represents the previous value of the outputs stored in D flip-flop. The truth table showing all the input combinations and respective output values is listed in Table 6.8. The presented 2-bit PIPO QCA layout in single layer is illustrated in Figure 6.15. It utilizes 56 cells, covers area of 0.05 μm^2 with a delay value of 0.75. The design is based on cell interaction approach, therefore the cost function results in value 3 only. The operation of circuit is also verified

TABLE 6.6
T flip-flop: power dissipation

Energy	Category Energy Dissipation Value in eV
Total energy dissipation (Sum Ebath) 8.04e-003	8.04e-003
Average energy dissipation per cycle (Avg Ebath) 7.31e-004	7.31e-004

TABLE 6.7
Comparison QCA T flip-flop

QCA T Flip-Flop Designs	Area Occupied (μm^2)	No. of Cells	Delay (Clock Cycles)	Cost Function
TFF-1 [Vetteth et al. (2003) Vetteth, Walus, Dimitrov, and Jullien]	0.1	92	1.25	36.25
TFF-2 [Torabi (2011)]	0.06	66	1.25	13.75
TFF-3 [Angizi et al. (2014) Angizi, Navi, Sayedsalehi, and Navin]	0.06	55	1.5	27
TFF-4 [Angizi et al. (2015) Angizi, Moaiyeri, Farrokhi, Navi, and Bagherzadeh]	0.06	46	1	18
TFF-5 [Majeed et al. (2019) Majeed, Alkaldy, bin Zainal, Nor et al.]	0.02	21	0.5	0.5
Proposed TFF	0.02	20	0.75	1.5

from its simulation results as illustrated in Figure 6.16. To measure the scalability of proposed DFF, 4-bit PIPO architecture have been designed as shown in Figure 6.17. The design utilizes four D flip-flop, 129 cells in an area of 0.14 μm^2 and produces output after 03 clock zones. The simulation results for 4-bit PIPO is demonstrated in Figure 6.18.

The proposed layout of 2-bit and 4-bit PIPO is also simulated on QCA Designer-E for power estimation. The simulation of 2-bit PIPO starts with 03 inputs and 02 outputs, takes 08 iterations to converge the initial steady state polarization. Whereas, 4-bit PIPO simulation initialize with 05 inputs and 04 outputs, requires 18 iterations to converge the initial steady state polarization. The total simulation time is found to be 33 seconds in 2-bit PIPO and 81 seconds in 4-bit PIPO respectively. The estimation of total and average energy per cycle has been tabulated in Table 6.9 and Table 6.10. It can be found in literature that few authors presented 4-bit PIPO designs with variations of complexities and design parameters. These are summarized and compared against performance measures area, cell count, delay, and cost function as tabulated in Table 6.11.

It can be observed from Table 6.11 that the proposed architecture is optimized in terms of area, delay, and cost-function. PIPO-2 has been reported with 114 cells but

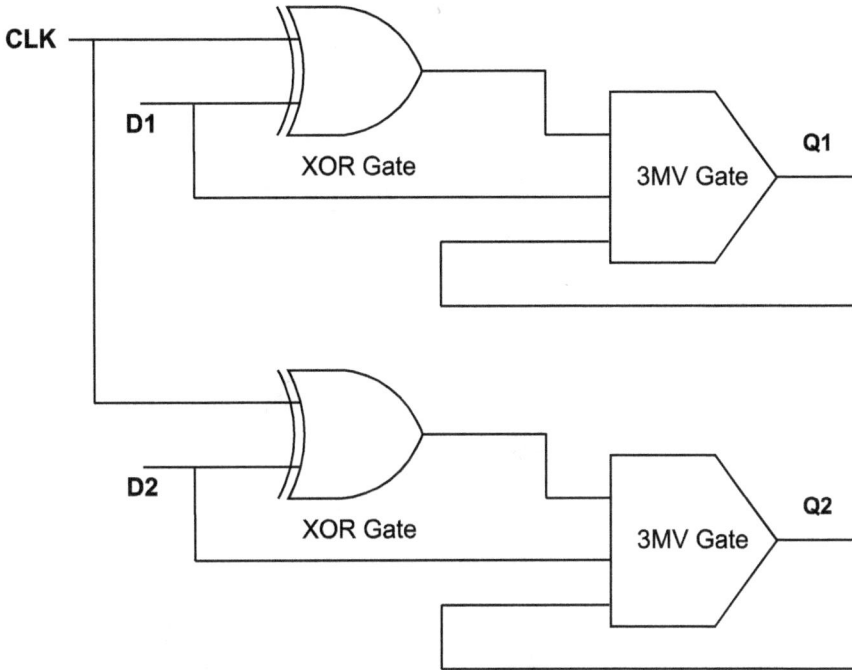

FIGURE 6.14 Logic diagram of 2-bit PIPO shift register.

TABLE 6.8
I/O behavior of 2-bit PIPO

Clock (CLK)	Input (D1)	Input (D2)	Output Q1	Output Q2
0	0	0	Q1(t-1)	Q2(t-1)
0	0	1	Q1(t-1)	Q2(t-1)
0	1	0	Q1(t-1)	Q2(t-1)
0	1	1	Q1(t-1)	Q2(t-1)
1	0	0	0	0
1	0	1	0	1
1	1	0	1	0
1	1	1	1	1

produces output after 04 clock zones whereas the proposed design of 4-bit PIPO takes only 03 clock zones to produce the required output. In addition, the prior designs utilizes twelve 3-input majority gate and four inverters which results in high complexity. Consequently, the designs are too costly with a value 148. PIPO-1 consists of 260 cells in an area of 1.67 μm^2, PIPO-3 utilizes 136 cells in 0.50 μm^2 area, and

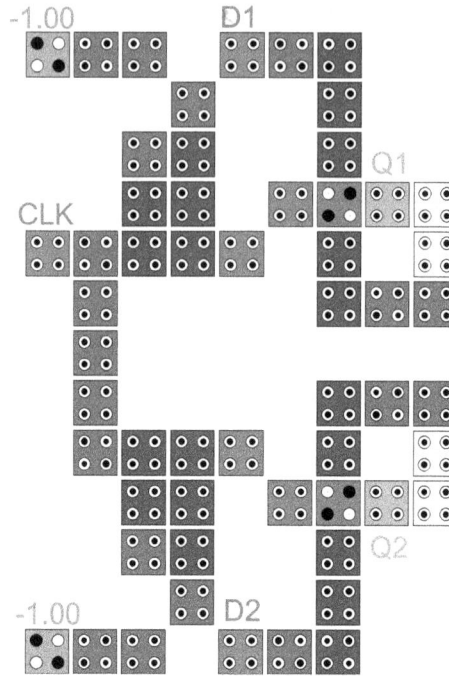

FIGURE 6.15   QCA layout of 2-bit PIPO.

PIPO-2 uses 114 cells and occupies 0.15 μm^2 area. The designs have been reported with similar speed and produce output after 04 clock zones.

6.3.4   COUNTER

A counter is a widely used sequential circuit designed using a combination of flip-flops which are connected in appropriate manner to count the number of clock signals applied as input to the circuit. Depending on the clock signal connection, a counter architecture is divided into two categories namely synchronous and asynchronous counter. If all the flip-flops are active at the same time with a common clock signal, the layout is termed as synchronous counter. Whereas, in case of asynchronous counter the input clock is applied to the first flip-flop and its output acts as clock input for the next successive flip-flop. Therefore, the overall delay of asynchronous counter is the sum of all individual flip-flop delay. However, the delay of synchronous counter is equal to the propagation delay of a single flip-flop. Hence, synchronous counters are advantageous and utilized widely in the designing of digital devices. It can be found in literature, that T flip-flop is used at primary level in the designing of synchronous counter architecture.

A N-Bit synchronous counter (SC) utilizes N number of flip-flops and counts from 0 to $2^{(N-1)}$. For example, a 2-bit counter counts from 0 to 3 and consists of two flip-flops. The logic level representation of 2-bit counter is illustrated in Figure 6.19. Here,

Simulation Results

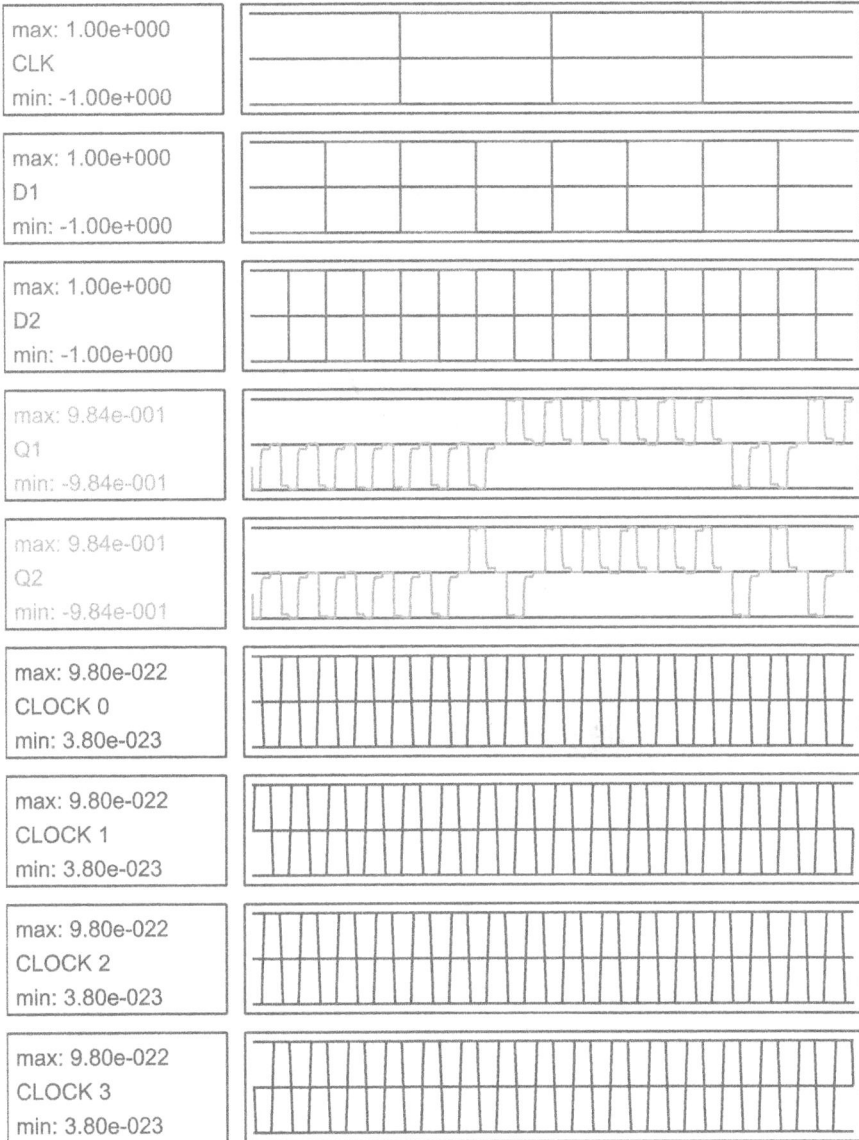

FIGURE 6.16   I/O waveform of 2-bit PIPO.

both flip-flops activate at the same time when the clock input signal is at HIGH logic state. The input of first T flip-flop $T1$ is fixed to logic "1" and its output is connected to second T flip-flop such as $T2 = Q1$. The truth table listing the next output state with respect to previous states has been tabulated in Table 6.12.

The proposed QCA architecture of 2-bit counter consists of 47 cells in an area of $0.05\ \mu m^2$ has been illustrated in Figure 6.20. The latency of layout is found as 1 which

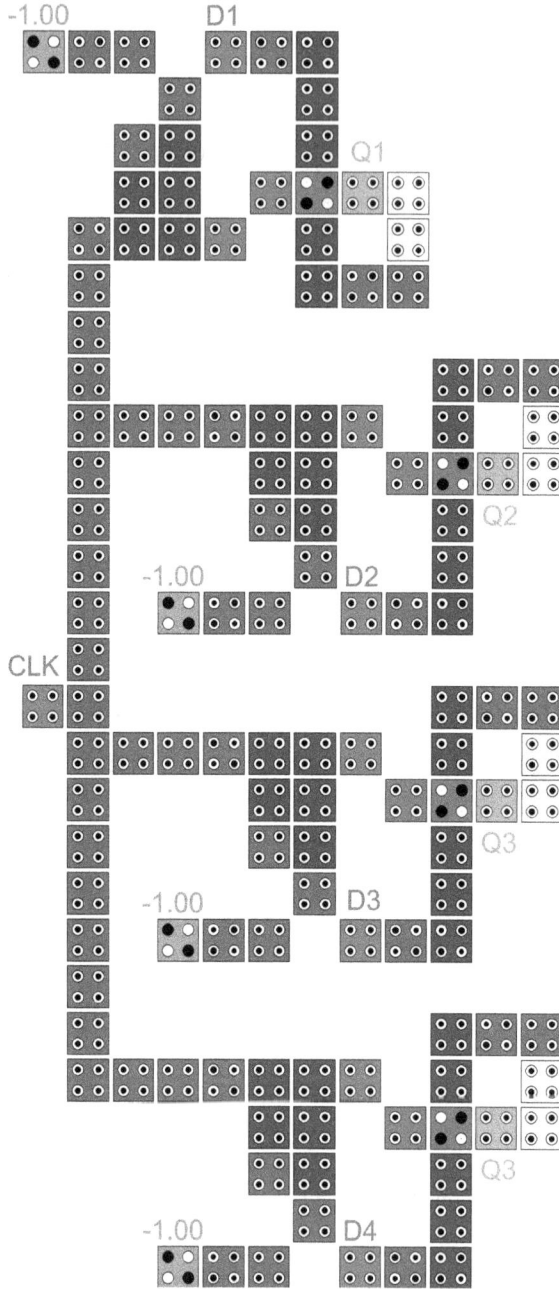

FIGURE 6.17 QCA layout of 4-bit PIPO.

Simulation Results

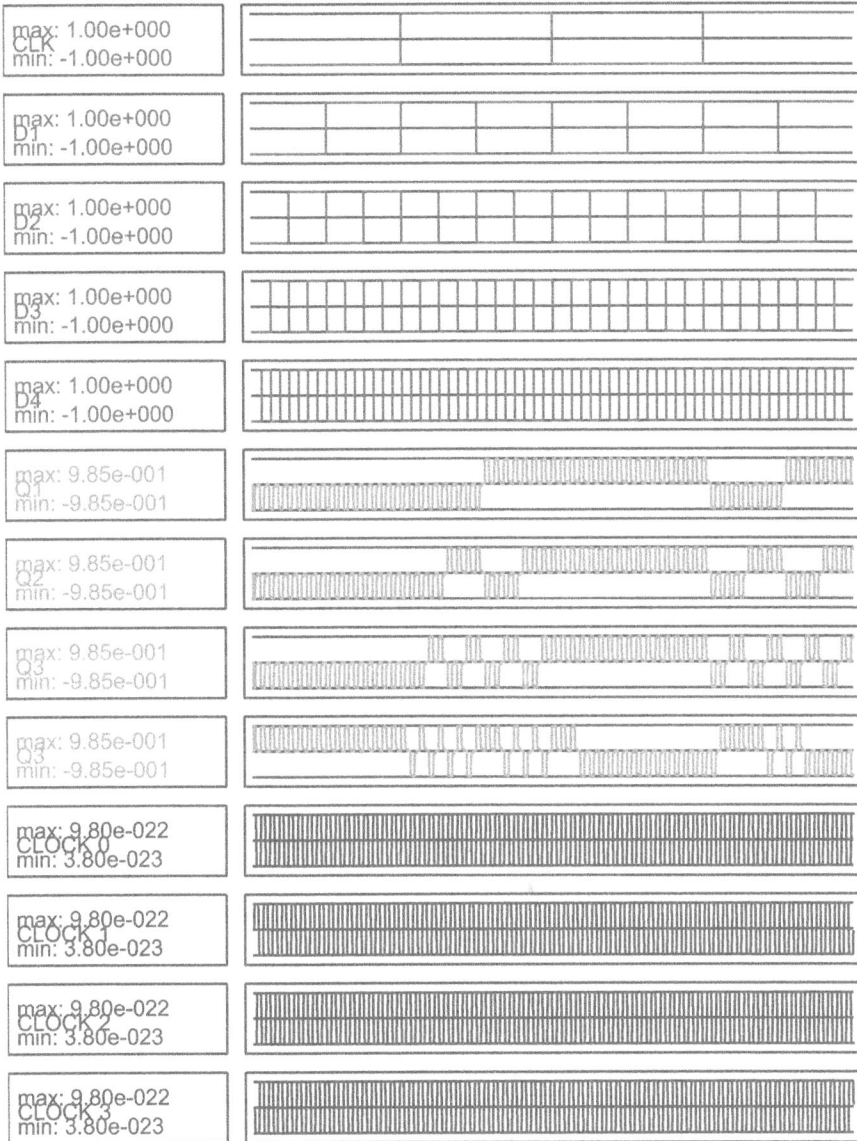

FIGURE 6.18   I/O waveform of 4-bit PIPO.

TABLE 6.9
2-bit PIPO: power dDissipation

Energy Category	Energy Dissipation Value in eV
Total energy dissipation (Sum Ebath)	2.42e-002
Average energy dissipation per cycle (Avg Ebath)	2.20e-003

TABLE 6.10
4-bit PIPO: Power Dissipation

Energy Category	Energy Dissipation Value in eV
Total energy dissipation (Sum Ebath)	4.21e-002
Average energy dissipation per cycle (Avg Ebath)	3.82e-003

TABLE 6.11
Comparison QCA 4-bit PIPO

QCA 4-bit PIPO Designs	Area Occupied (μm^2)	No. of Cells	Delay (Clock Cycles)	Cost Function
PIPO-1 [Purkayastha et al. (2018) Purkayastha, De, and Chattopadhyay]	1.67	260	1	148
PIPO-2 [Nafees et al. (2022) Nafees, Ahmed, Kakkar, Bahar, Wahid, and Otsuki]	0.15	114	1	148
PIPO-3 [Jeon (2020)]	0.5	136	1	148
Proposed 2-bit PIPO	0.05	56	0.75	3
Proposed 4-bit PIPO	0.14	129	0.75	12

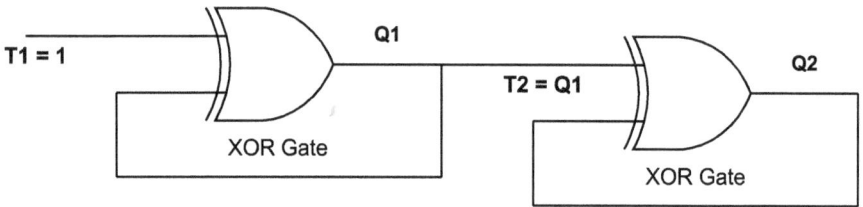

FIGURE 6.19   Logic diagram of 2-bit counter.

TABLE 6.12
I/O behavior of 2-bit counter

Present State		Next State	
Q1(t-1)	Q2(t-1)	Q1(t)	Q2(t)
0	0	0	1
0	1	1	0
1	0	1	1
1	1	0	0

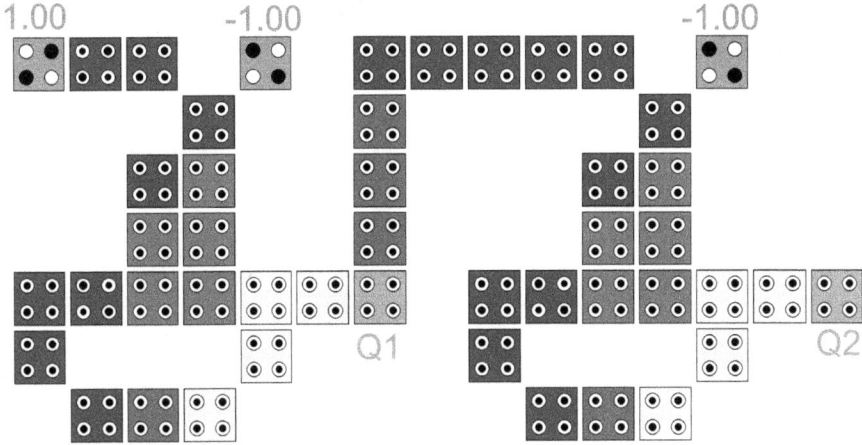

FIGURE 6.20   QCA layout of 2-bit counter.

shows that the output is generated after 04 clock zones. The simulation results that verifies circuit operation is shown in Figure 6.21.

The power estimation is done using QCA designer-E tool which initiates with 00 inputs and 02 output, takes total simulation time as 28 seconds. The iterations required to converge the initial steady state polarization is found to be 40. The total and average energy per cycle findings in simulation have been listed in Table 6.13. Several counter designs of length 2-bit have been reported in literature with variations of design parameters and complexities. These are tabulated and compare conversely in terms of performance metrics Area, cell count, delay and cost function as tabulated in Table 6.14.

It can be observed from Table 6.14 that the proposed architecture proves itself superior over prior reported designs in all performance metrics. SC-1 architecture is found to be complex as it uses fourteen 3-input majority gate with six inverters and two crossover. Consequently, the design becomes costly with a value 618. It consists of 328 cells in an area of 0.62 μm^2 and takes 12 clock zones to produce the required output. Utilizing ten 3-input majority gate and six inverters with one crossover, SC-2 has been reported.

The design consumes 240 cells in 0.26 μm^2 area and produces output after 08 clock zones. Two design SC-3 and SC-4 have been reported with same speed and cell count difference by a value 1. However, SC-4 is less complex as compared to SC-3 as the former utilized less number of 3-input majority gate which results in low cost-function to value 90. In addition, the area occupied by SC-3 and SC-4 is 0.22 μm^2 and 0.17 μm^2 respectively. Utilizing three 3-input majority gate SC-5 have been reported with 80 cells in an area of 0.09 μm^2. The cost is only 18 and produces output after 04 clock zones. SC-6 has been reported as a high speed counter design since only 02 clock zones are required to produce the desired output. It takes 78 cells and occupies 0.033 μm^2 area in single layer with a cost value of 14.

Simulation Results

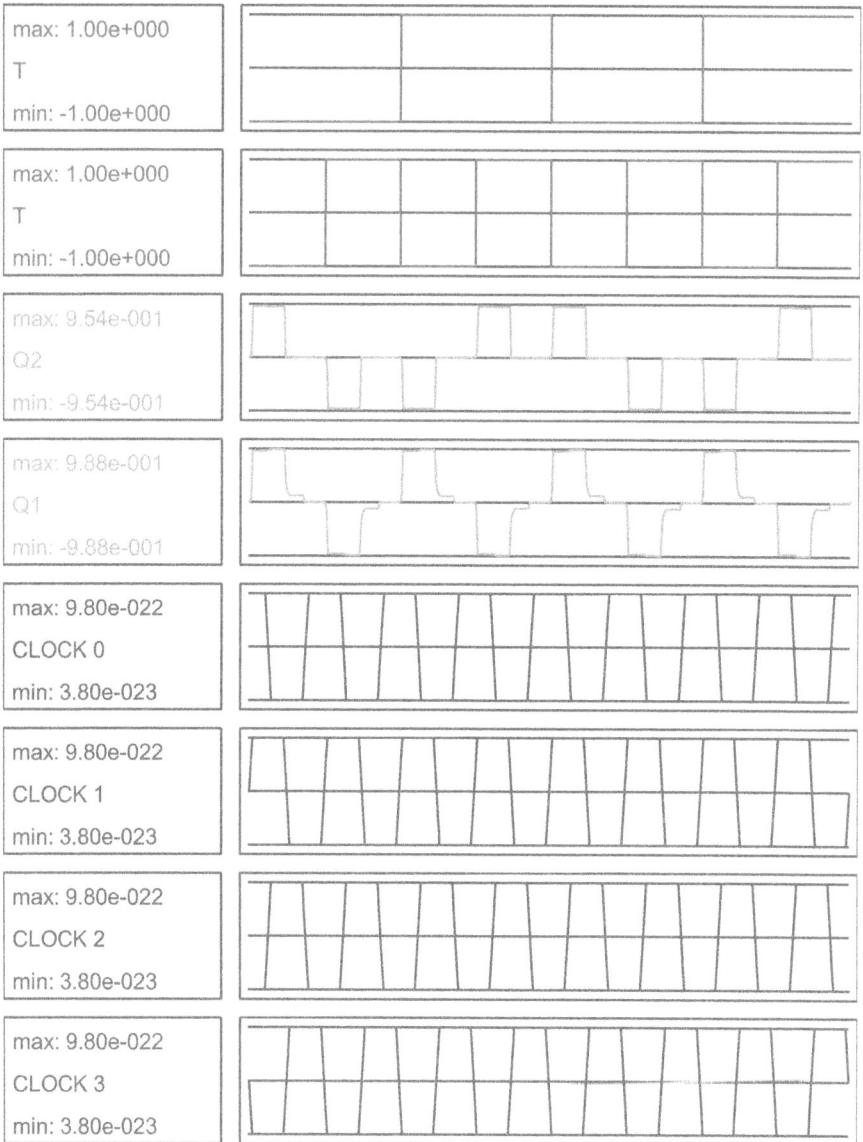

max: 1.00e+000 T min: -1.00e+000	
max: 1.00e+000 T min: -1.00e+000	
max: 9.54e-001 Q2 min: -9.54e-001	
max: 9.88e-001 Q1 min: -9.88e-001	
max: 9.80e-022 CLOCK 0 min: 3.80e-023	
max: 9.80e-022 CLOCK 1 min: 3.80e-023	
max: 9.80e-022 CLOCK 2 min: 3.80e-023	
max: 9.80e-022 CLOCK 3 min: 3.80e-023	

FIGURE 6.21 I/O Waveform of 2-bit counter.

TABLE 6.13
2-bit counter: power dissipation

Energy Category	Energy Dissipation Value in eV
Total energy dissipation (Sum Ebath)	1.96e-002
Average energy dissipation per cycle (Avg Ebath)	1.79e-003

TABLE 6.14
Comparison QCA 2-bit counter

QCA 2-bit Counter Designs	Area Occupied (μm2)	No. of Cells	Delay (Clock Cycles)	Cost Function
SC-1 [Yang et al. (2010) Yang, Cai, Zhao, and Zhang]	0.62	328	3	618
SC-2 [Sheikhfaal et al. (2015) Sheikhfaal, Navi, Angizi, and Navin]	0.26	240	2	214
SC-3 [Angizi et al. (2015) Angizi, Moaiyeri, Farrokhi, Navi, and Bagherzadeh]	0.22	141	2.25	191.25
SC-4 [Bhavani and Alinvinisha (2015)]	0.17	140	2.25	90
SC-5 [Majeed et al. (2019) Majeed, Alkaldy, bin Zainal, Nor et al.]	0.09	80	2	18
SC-6 [Sangsefidi et al. (2018) Sangsefidi, Abedi, Yoosefi, and Karimpour]	0.06	78	0.5	14
Proposed	0.05	47	1	3

6.4 POWER ESTIMATION USING QCA-PRO

The total and average power per cycle during simulation for each proposed design have been computed using QCA Designer-E tool. Another popular software termed as QCA- Pro is also utilized to measure non-adiabatic switching power with polarization loss at a fixed operating temperature level with value 2 K. In addition, measurement carried out at three different tunneling energy levels termed as 0.5Ek, 1.0Ek, and 1.5Ek [Srivastava et al. (2011) Srivastava, Asthana, Bhanja, and Sarkar]. It is a probabilistic modeling tool which uses the fast approximation technique to measure cells errors. The tool measures average leakage and switching energy which adds together to compute total leakage energy. This section further describes the power estimation calculation of proposed D and T flip-flop, register and counter circuit along with thermal power dissipation map. The map indicates dissipated heat of individual cell which is directly proportional to its darkness and the transparent cell represents input.

6.4.1 COMPUTATION FOR D FLIP-FLOP

The power estimation of D flip-flop requires the development of switching vector matrix having size 4 × 2 and input-output vector matrix with size 4 × 3. Here the switching matrix is a combination of input variables only whereas I/O vector set represent both input and output variables. The average leakage and switching

TABLE 6.15
Energy dissipation analysis of D flip-flop

Energy Dissipation (meV)	0.5Ek	1.0Ek	1.5Ek
Avg. Switching	0.013	0.01	0.009
Avg. leakage	0.012	0.03	0.051
Total	0.025	0.04	0.06

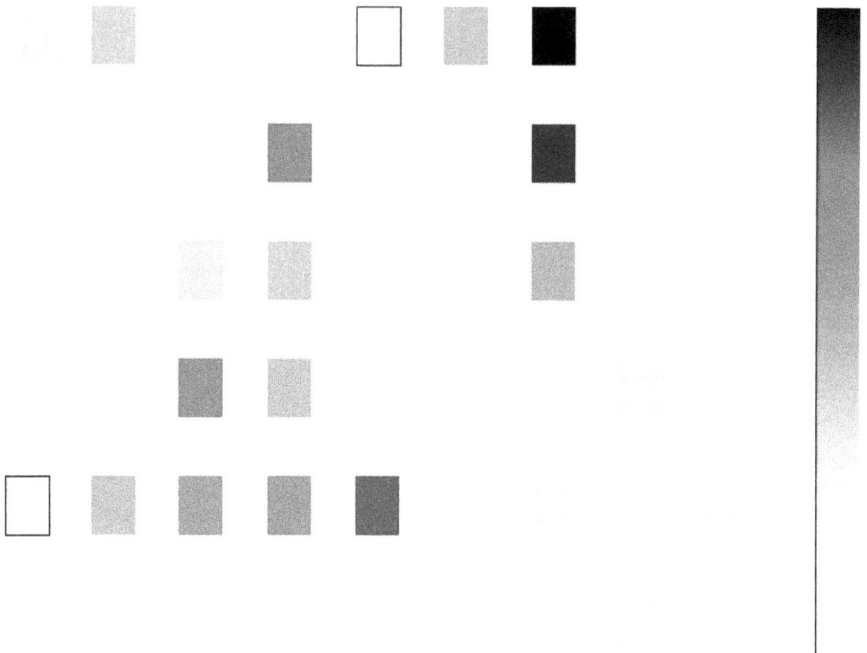

FIGURE 6.22 Thermal energy dissipation map of D flip-flop at 0.5Ek.

energies with total dissipation energy for different energy levels has been tabulated in Table 6.15.

The thermal power dissipation map at 0.5Ek, 1.0Ek, and 1.5Ek is illustrated in Figure 6.22, Figure 6.23 and Figure 6.24 respectively.

6.4.2 COMPUTATION FOR T FLIP-FLOP

The power estimation of T flip-flop requires the development of switching vector matrix having size 4 × 2 and input-output vector matrix with size 4 × 3. Here the

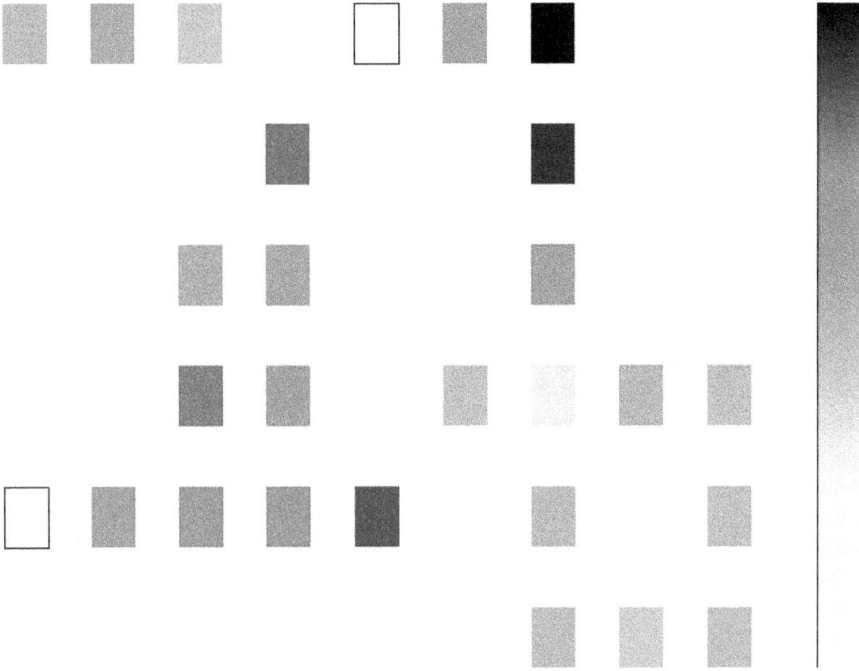

FIGURE 6.23 Thermal energy dissipation map of D flip-flop at 1.0Ek.

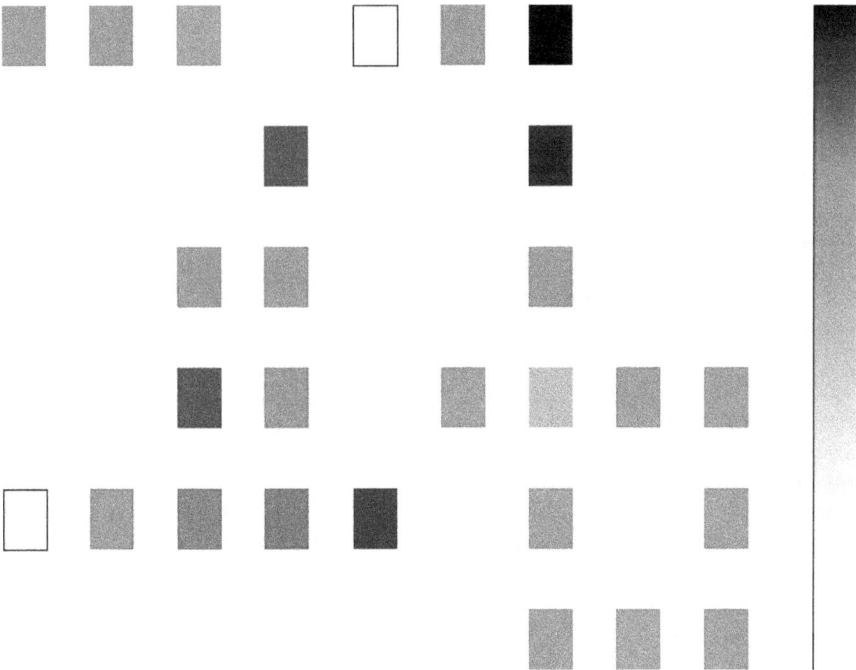

FIGURE 6.24 Thermal energy dissipation map of D flip-flop at 1.5Ek.

TABLE 6.16
Energy dissipation analysis of T flip-flop

Energy Dissipation (meV)	0.5Ek	1.0Ek	1.5Ek
Avg. Switching	0.002	0.001	0.001
Avg. leakage	0.009	0.022	0.037
Total	0.011	0.023	0.038

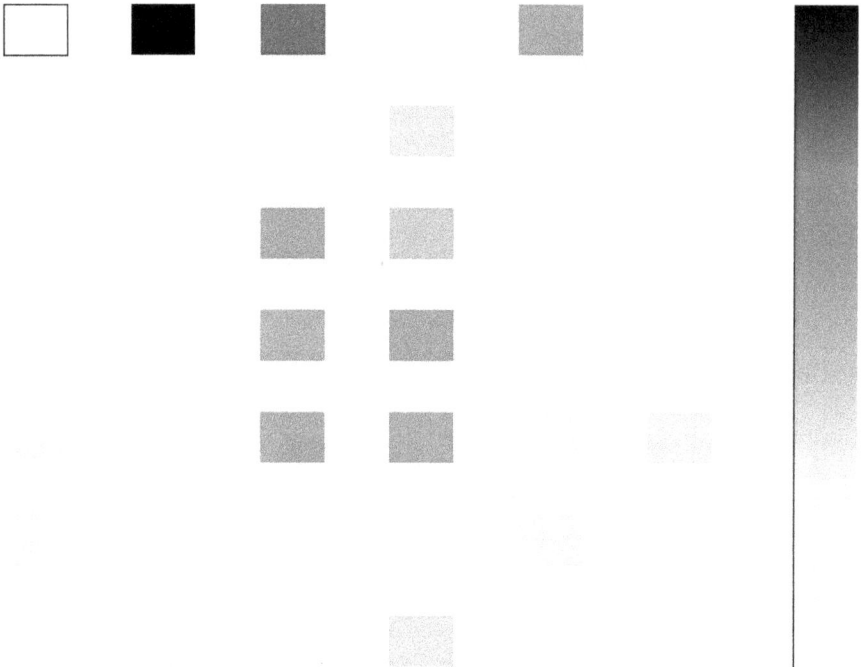

FIGURE 6.25   Thermal energy dissipation map of T flip-flop at 0.5Ek.

switching matrix is a combination of input variables only whereas I/O vector set represent both input and output variables. The average leakage and switching energies with total dissipation energy for different energy levels has been tabulated in Table 6.16.

The thermal power dissipation map at 0.5Ek, 1.0Ek, and 1.5Ek is illustrated in Figure 6.25, Figure 6.26 and Figure 6.27 respectively.

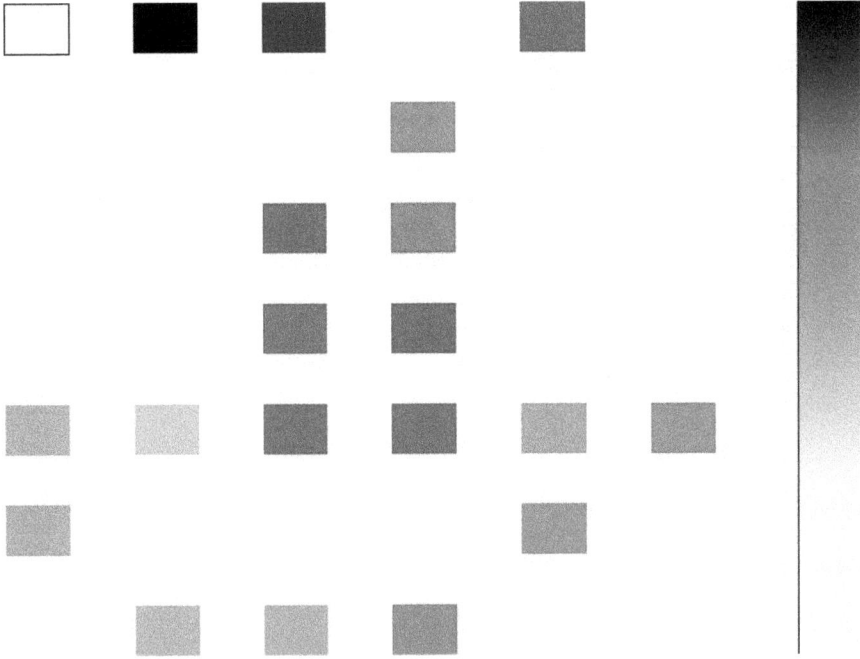

FIGURE 6.26   Thermal energy dissipation map of T flip-flop at 1.0Ek.

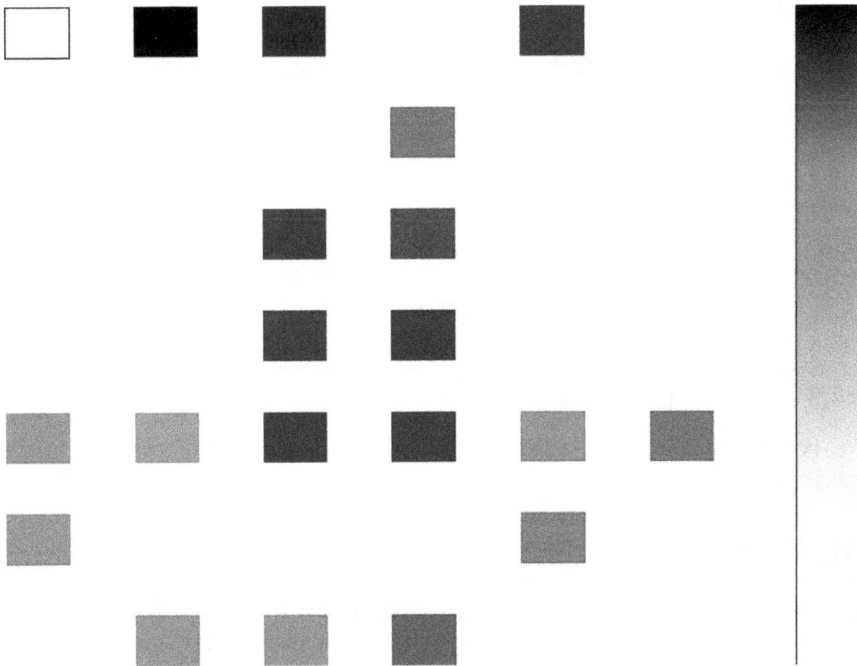

FIGURE 6.27   Thermal energy dissipation map of T flip-flop at 1.5Ek.

TABLE 6.17
Energy dissipation analysis of 2-bit PIPO

Energy Dissipation (meV)	0.5Ek	1.0Ek	1.5Ek
Avg. Switching	0.029	0.023	0.012
Avg. leakage	0.024	0.062	0.103
Total	0.053	0.085	0.115

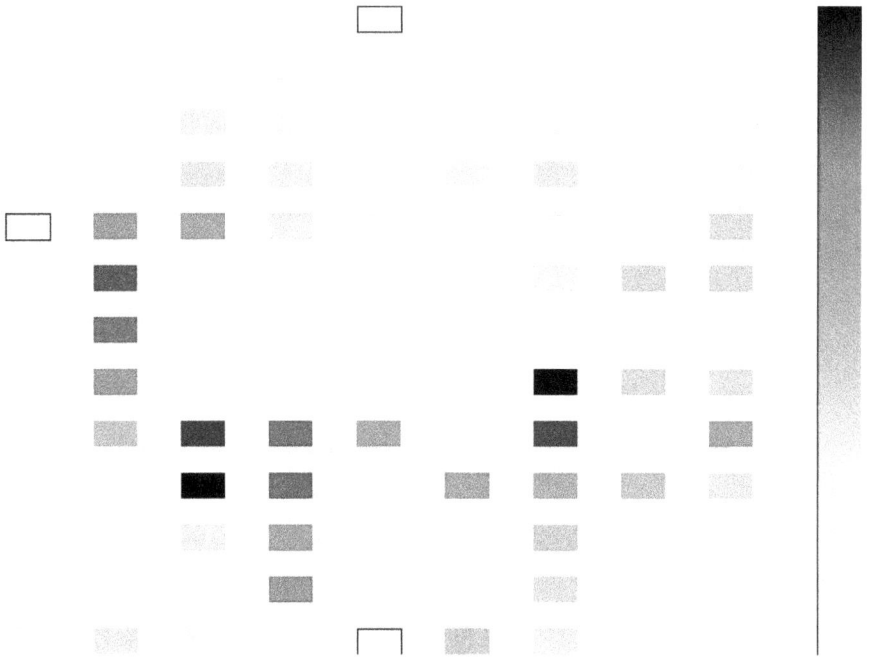

FIGURE 6.28   Thermal energy dissipation map of 2-bit PIPO at 0.5Ek.

6.4.3   Computation for 2-bit PIPO

The power estimation of 2-bit PIPO requires the development of switching vector matrix having size 8×5 and input-output vector matrix with size 8×3. Here the switching matrix is a combination of input variables only whereas I/O vector set represent both input and output variables. The average leakage and switching energies with total dissipation energy for different energy levels has been tabulated in Table 6.17. The thermal power dissipation map at 0.5Ek, 1.0Ek, and 1.5Ek is illustrated in Figure 6.28, Figure 6.29 and Figure 6.30 respectively.

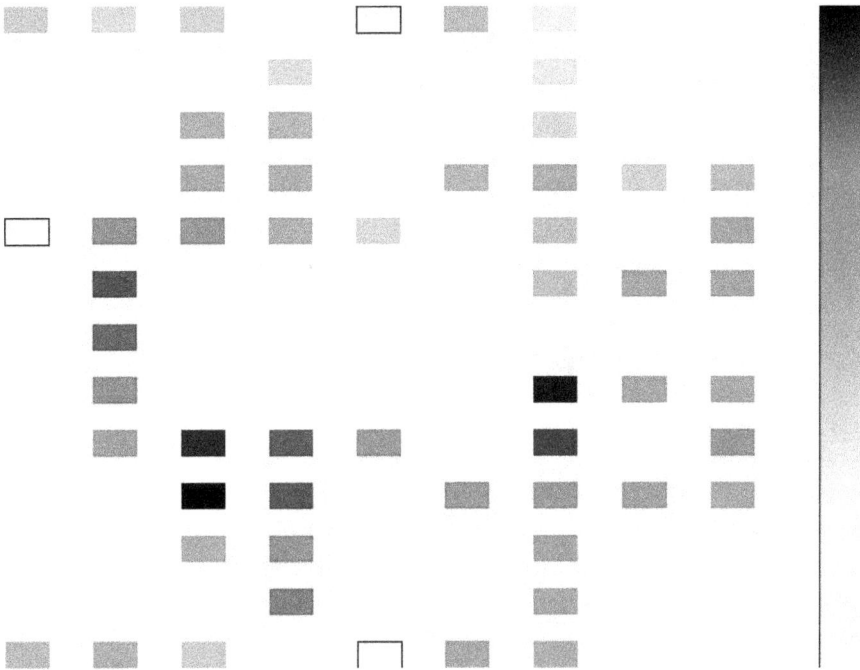

FIGURE 6.29   Thermal energy dissipation map of 2-bit PIPO at 1.0Ek.

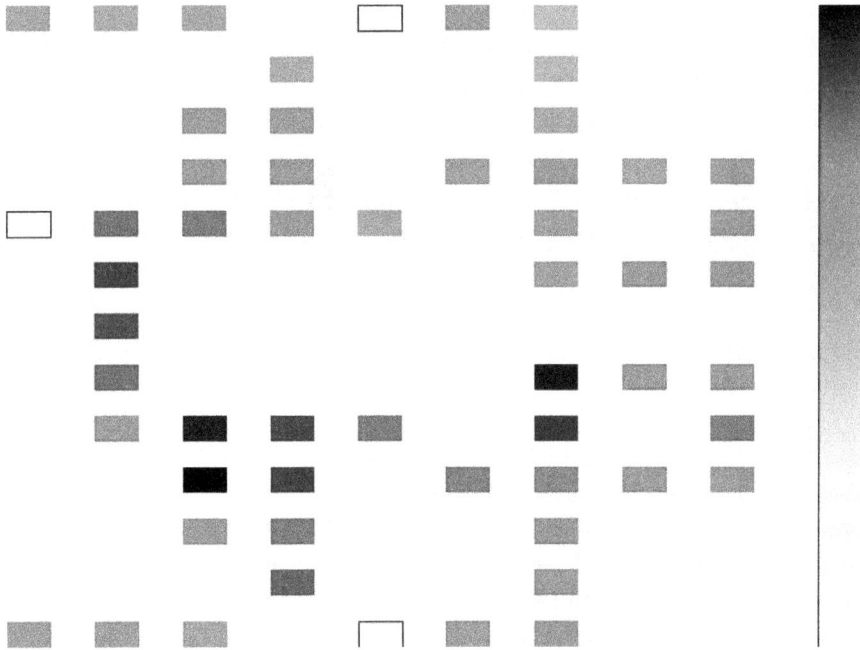

FIGURE 6.30   Thermal energy dissipation map of 2-bit PIPO at 1.5Ek.

TABLE 6.18
Energy dissipation analysis of 2-bit counter

Energy Dissipation (meV)	0.5Ek	1.0Ek	1.5Ek
Avg. Switching	0.000	0.000	0.000
Avg. leakage	0.017	0.049	0.086
Total	0.017	0.049	0.086

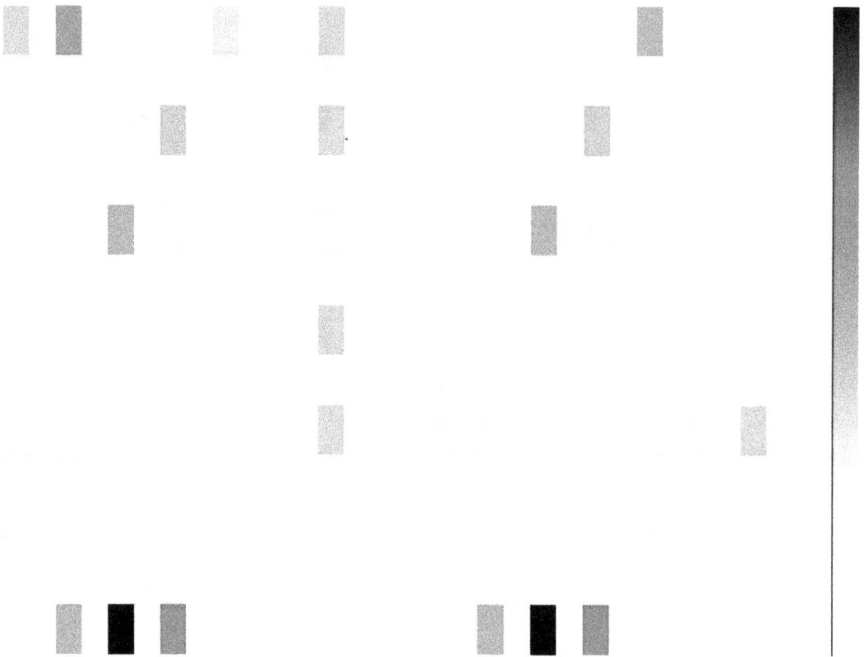

FIGURE 6.31   Thermal energy dissipation map of 2-bit counter at 0.5Ek.

6.4.4   COMPUTATION FOR 2-BIT COUNTER

The power estimation of 2-bit counter requires the development of switching vector matrix having size 2×4 and input-output vector matrix with size 4×4. Here the switching matrix is a combination of input variables only whereas I/O vector set represent both input and output variables. The average leakage and switching energies with total dissipation energy for different energy levels has been tabulated in Table 6.18. The thermal power dissipation map at 0.5Ek, 1.0Ek and 1.5Ek is illustrated in Figure 6.31, Figure 6.32 and Figure 6.33 respectively.

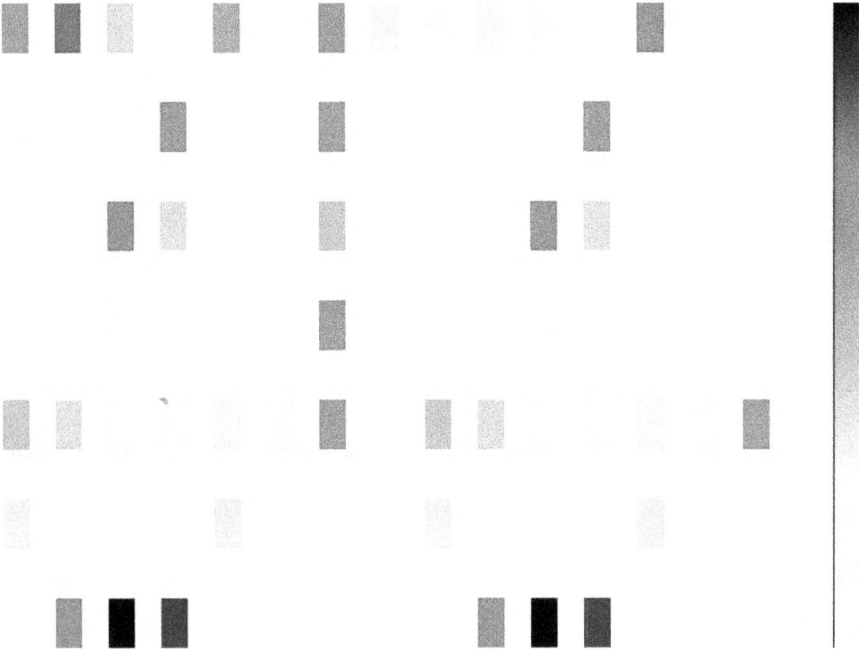

FIGURE 6.32 Thermal energy dissipation map of 2-bit counter at 1.0Ek.

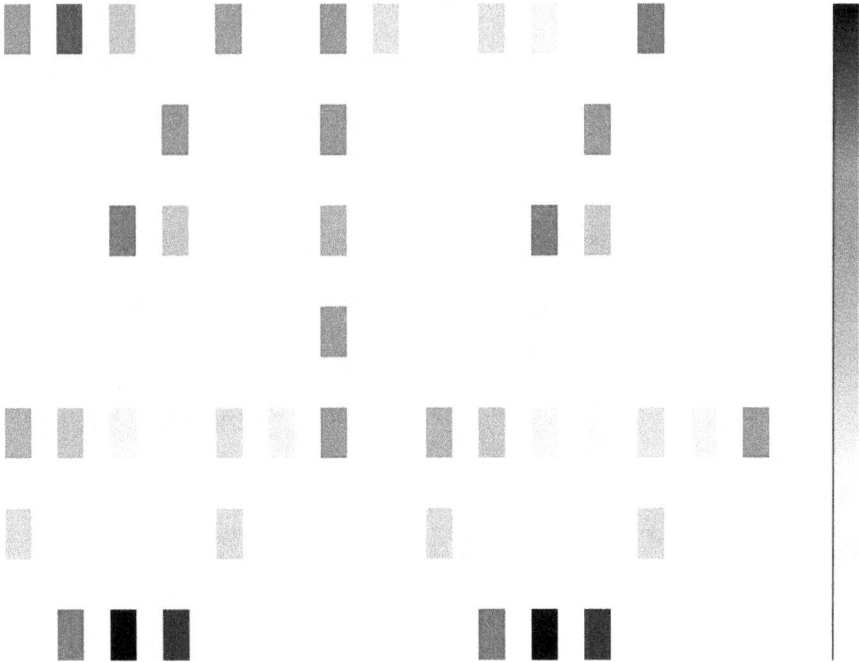

FIGURE 6.33 Thermal energy dissipation map of 2-bit counter at 1.5Ek.

6.5 CONCLUSION

Among several alternative to CMOS technology, QCA proved itself the best replacement and hence used in the present work. It can be seen in literature, that flip-flop design is being used at primary level in the designing of sequential circuit. Several flip-flop designs are reported previously which need improvement due to higher cell count, area, delay, and cost. Therefore, this chapter proposed new improved designs of D and T flip-flop at primary level which are further utilized to build other sequential circuits such as register and counter of length 2-bit. The new designs are compared with previously reported designs and are found efficient somewhere in area, cell count, delay, and cost-function. QCA designer 2.0.1 has been utilized for designing and to simulate circuits in QCA technology. The total and average power per cycle of the designed circuit is computed using QCA Designer-E tool. However, QCAPro is used to measure switching power loss and polarization error. The proposed designs of flip-flops found to be scalable at larger level and can be used in future to design a larger size register, memory, and counter circuits.

REFERENCES

M. Abutaleb, "Robust and efficient quantum-dot cellular automata synchronous counters," *Microelectronics Journal*, vol. 61, pp. 6–14, 2017.

H. Alamdar, G. Ardeshir, and M. Gholami, "Phase-frequency detector in qca nanotechnology using novel flip-flop with reset terminal," *International Nano Letters*, vol. 10, no. 2, pp. 111–118, 2020.

——, "Using universal nand- nor-inverter gate to design d-latch and d flip-flop in quantum-dot cellular automata nanotechnology," *International Journal of Engineering*, vol. 34, no. 7, pp. 1710–1717, 2021.

S. Angizi, K. Navi, S. Sayedsalehi, and A. H. Navin, "Efficient quantum dot cellular automata memory architectures based on the new wiring approach," *Journal of Computational and Theoretical Nanoscience*, vol. 11, no. 11, pp. 2318–2328, 2014.

S. Angizi, M. H. Moaiyeri, S. Farrokhi, K. Navi, and N. Bagherzadeh, "Designing quantum-dot cellular automata counters with energy consumption analysis," *Microprocessors and Microsystems*, vol. 39, no. 7, pp. 512–520, 2015.

K. S. Bhavani and V. Alinvinisha, "Utilization of QCA based t flip flop to design counters," in *2015 International Conference on Innovations in Information, Embedded and Communication Systems (ICIIECS)*. IEEE, 2015, pp. 1–6.

R. Chakrabarty, D. K. Mahato, A. Banerjee, S. Choudhuri, M. Dey, and N. K. Mandal, "A novel design of flip-flop circuits using quantum dot cellular automata (QCA)," in *2018 IEEE 8th Annual Computing and Communication Workshop and Conference (CCWC)*. IEEE, 2018, pp. 408–414.

H. M. Gaur and A. Singh, "Design of reversible circuits with high testability," *Electronics Letters*, vol. 52, no. 13, pp. 1102–1104, 2016.

H. M. Gaur, A. K. Singh, and U. Ghanekar, "A review on online testability for reversible logic," *Procedia Computer Science*, vol. 70, pp. 384–391, 2015.

H. M. Gaur, T. Sasamal, A. Singh, A. Mohan, and D. Pradhan, "Reversible logic: An introduction," in *Design and Testing of Reversible Logic*. Springer, 2020, pp. 3–18.

M. Goswami, B. Kumar, H. Tibrewal, and S. Mazumdar, "Efficient realization of digital logic circuit using QCA multiplexer," in *2014 2nd International Conference on Business and Information Management (ICBIM)*. IEEE, 2014, pp. 165–170.

S. Hashemi and K. Navi, "New robust QCA d flip flop and memory structures," *Microelectronics Journal*, vol. 43, no. 12, pp. 929–940, 2012.

V. Jain, D. K. Sharma, and H. M. Gaur, "Area and energy optimized multilayer QCA-based 4n-bit scalable multiplier (m4n- mul)," *The European Physical Journal Plus*, vol. 137, no. 11, p. 1281, 2022.

J.-C. Jeon, "Low-complexity QCA universal shift register design using multiplexer and d flip-flop based on electronic correlations," *The Journal of Super-computing*, vol. 76, no. 8, pp. 6438–6452, 2020.

C. S. Lent and P. D. Tougaw, "A device architecture for computing with quantum dots," *Proceedings of the IEEE*, vol. 85, no. 4, pp. 541–557, 1997.

W. Liu, L. Lu, M. ONeill, and E. E. Swartzlander, "A first step toward cost functions for quantum-dot cellular automata designs," *IEEE Transactions on Nanotechnology*, vol. 13, no. 3, pp. 476–487, 2014.

C. Mack, "The multiple lives of moore's law," *IEEE Spectrum*, vol. 52, no. 4, pp. 31–31, 2015.

A. H. Majeed, E. Alkaldy, M. S. bin Zainal, B. M. Nor *et al.*, "Synchronous counter design using novel level sensitive t-ff in QCA technology," *Journal of Low Power Electronics and Applications*, vol. 9, no. 3, p. 27, 2019.

N. Nafees, S. Ahmed, V. Kakkar, A. N. Bahar, K. A. Wahid, and A. Otsuki, "QCA-based pipo and sipo shift registers using cost-optimized and energy-efficient d flip flop," *Electronics*, vol. 11, no. 19, p. 3237, 2022.

T. Purkayastha, D. De, and T. Chattopadhyay, "Universal shift register implementation using quantum dot cellular automata," *Ain Shams Engineering Journal*, vol. 9, no. 2, pp. 291–310, 2018.

M. Sangsefidi, D. Abedi, E. Yoosefi, and M. Karimpour, "High speed and low cost synchronous counter design in quantum-dot cellular automata," *Microelectronics Journal*, vol. 73, pp. 1–11, 2018.

T. N. Sasamal, A. K. Singh, and U. Ghanekar, "Design of QCA-based d flip flop and memory cell using rotated majority gate," in *Smart Innovations in Communication and Computational Sciences*. Springer, 2019, pp. 233–247.

S. Sheikhfaal, K. Navi, S. Angizi, and A. Habibizad, "Designing high speed sequential circuits by quantum-dot cellular automata: Memory cell and counter study," *Quantum Matter*, vol. 4, no. 2, pp. 190–197, 2015.

S. Srivastava, A. Asthana, S. Bhanja, and S. Sarkar, "Qcapro-an error-power estimation tool for qca circuit design," in *2011 IEEE International Symposium of Circuits and Systems (ISCAS)*. IEEE, 2011, pp. 2377–2380.

M. Torabi, "A new architecture for t flip flop using quantum-dot cellular automata," in *2011 3rd Asia Symposium on Quality Electronic Design (ASQED)*. IEEE, 2011, pp. 296–300.

F. S. Torres, R. Wille, P. Niemann, and R. Drechsler, "An energy-aware model for the logic synthesis of quantum-dot cellular automata," *IEEE Transactions on Computer-Aided Design of Integrated Circuits and Systems*, vol. 37, no. 12, pp. 3031–3041, 2018.

P. D. Tougaw and C. S. Lent, "Logical devices implemented using quantum cellular automata," *Journal of Applied Physics*, vol. 75, no. 3, pp. 1818–1825, 1994.

——, "Dynamic behavior of quantum cellular automata," *Journal of Applied Physics*, vol. 80, no. 8, pp. 4722–4736, 1996.

P. D. Tougaw, W. Porod, G. H. Bernstein, "Quantum Cellular Automata," *Nanotechnology*, vol. 4 pp. 49–57, 1993.

A. Vetteth, K. Walus, V. S. Dimitrov, and G. A. Jullien, "Quantum-dot cellular automata of flip-flops," *ATIPS Laboratory*, vol. 2500, pp. 1–5, 2003.

X. Yang, L. Cai, X. Zhao, and N. Zhang, "Design and simulation of sequential circuits in quantum-dot cellular automata: Falling edge-triggered flip-flop and counter study," *Microelectronics Journal*, vol. 41, no. 1, pp. 56–63, 2010.

7 QCA-based Designs of Majority Gates, Flip-Flops and Polar Encoders

Aibin Yan, Aoran Cao, Runqi Liu
and Xuehua Li

7.1 INTRODUCTION

7.1.1 BACKGROUND

The exponential shrinking of transistor feature sizes causes severe challenges for energy dissipation and the manufacturing of CMOS technologies [1]. With the decrease of transistor feature sizes in CMOS technology, devices are more susceptible to high leakage current, high power density, and thermal noise. The performance bottleneck of transistors will have a huge impact on the VLSI industry. In order to deal with these upcoming challenges, high-performance next generation electronic devices have become a research hotspot. At present, researchers have proposed many nanoelectronic devices, including carbon nanotube transistors, silicon nanowires, spin transistors, single electron transistors, resonant tunneling devices, and Quantum-Dot Cellular Automata. Quantum-Dot Cellular Automata (QCA) technology is one of the most important nanotechnologies, and it is currently being treated as an alternative to Complementary Metal Oxide Semiconductor (CMOS) [3–4]. QCA technology has the advantages of ultra-high density, fast switching speed in the terahertz frequency range, and ultra-low power consumption. The favorable characteristics of QCA make it possible to break through the bottleneck of traditional circuits and implement more efficient circuit performance.

The following materials and methods have been used to create QCA devices: (1) semiconductor material scheme [5]; (2) ferromagnetic material scheme [6]; (3) molecular material scheme [7]; and (4) strategy of tunnel junction produced from metal material [8]. Molecular QCA cells and magnetic QCA cells can work stably at room temperature, which provides a reliable premise for large-scale QCA circuits that can operate reliably at room temperature in the later stage.

7.1.2 QCA BASICS

A four-dot QCA cell is the most basic component unit of QCA. There are four quantum dots, two free electrons in a four-dot QCA cell. There is a tunnel junction

DOI: 10.1201/9781003361633-7

between each quantum dot. Two free electrons can move freely among the four quantum dots in a four-dot QCA cell. The polarization P of a four-dot QCA cell, which can measure the extent of electronic charge distribution among four dots, can be defined by Formula 1. ρ_i represents the electronic charge in each quantum dot of each cell. A four-dot QCA cell can reach two stable states due to electrostatic interaction. The two polarization states of a QCA cell can be denoted as p = +1 and p = −1, which can represent the binary '1' and binary '0', respectively. Four-dot QCA cells can be divided into normal cells and rotated cells. Figure 7.1 shows the structure and type of four-dot QCA cells. Figure 7.1(a) shows the structure of four-dot QCA cells. Figure 7.1(b) shows the normal four-dot QCA cells and rotated four-dot QCA cells. The QCA cell mentioned in this chapter refers to four-dot QCA cells.

$$P = \frac{(\rho_1 + \rho_3) - (\rho_2 + \rho_4)}{\rho_1 + \rho_2 + \rho_3 + \rho_4} \tag{1}$$

A QCA circuit is separated by different clock zones so that all QCA cells in each zone are controlled by the same clock signal and perform specific operations. All QCA circuits require accurate clocking to manage the flow of information and the clocking can provide the necessary power to operate the QCA cells. To drive the input to the output to obtain a value, signals must pass through four clock zones. Figure 7.2 shows the QCA clock. Switch, hold, release, and relax are four clock phases in a clock. In the switch phase, the cell enters a polarized state according to the state of the drive cell. In the hold stage, the maximum polarity of cells remained unchanged. During the release phase, the cell begins to lose polarity. In the relax phase, the cell is in a non-polarized state [9]. These clocks are 90 degrees out of phase so as to ensure the correct transmission of signals. These clock zones support sequential computing, which means that when one zone is processing data, the previous zone must keep the results. Switch, hold, release, and relax are the four stages of each signal clocking. The cells of the four clocks are represented by four different colors.

The majority gate and inverter are critical components in QCA-based circuit design. Figure 7.3 shows the widely used logic gates in QCA. Figure 7.3(a) shows an inverter, Figure 7.3(b) and Figure 7.3(c) show two 3-input majority gate. The inverter can reverse an input signal and the 3-input majority gate can output a signal that follows the majority rule. The following equation describes the 3-input majority gate function [3]:

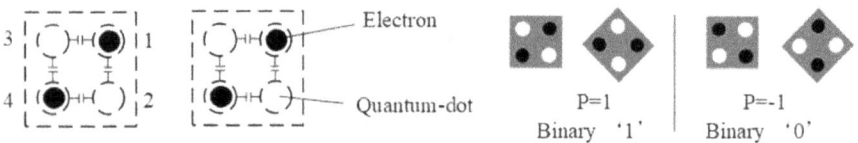

FIGURE 7.1 Structure and type of four-dot QCA cells. (a) Structure and (b) Two types of QCA cells.

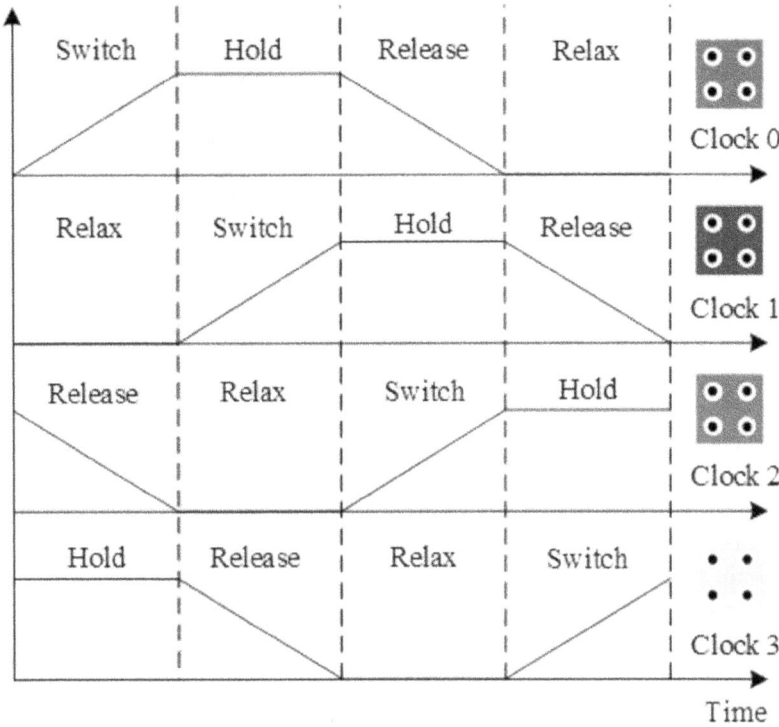

FIGURE 7.2 QCA clock.

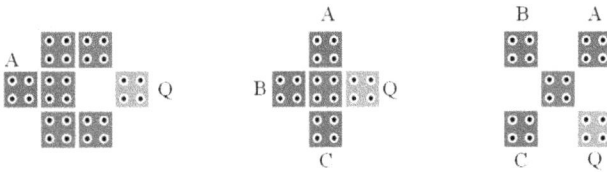

FIGURE 7.3 Widely used logic gates in QCA.

$$M(A,B,C) = AB + BC + AC \qquad (2)$$

Figure 7.4 shows several different crossovers. Multi-layer crossover uses the way of multi-layer wiring to implement the cross-transmission of signals. The disadvantage is that it is difficult to manufacture. Coplanar crossovers can be implemented by rotated QCA cells and normal QCA cells. When rotated QCA cells and normal QCA cells transmit signals, the Coulomb force between cells does not interfere with neighboring cells because the electron configuration of the two cells is orthogonal [10]. Because crossovers using rotated cells are irregular, Shin et al. proposed a new coplanar cross-wire using different clock zones (using clock zones 1 and 3 or clock

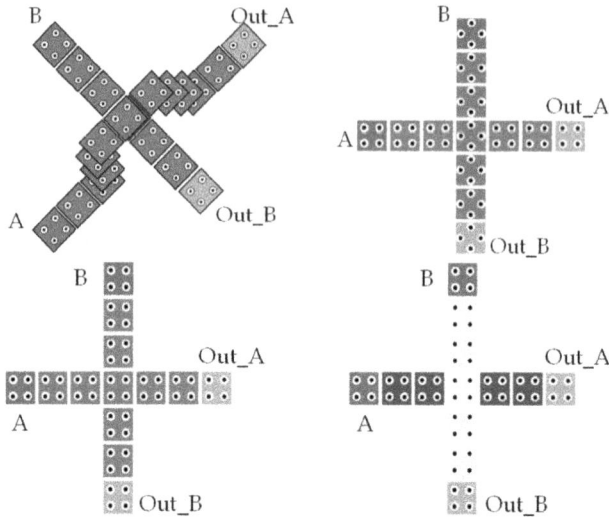

FIGURE 7.4 Crossovers. (a) Multilayer crossover, (b) Coplanar crossover implemented by rotating QCA cells in [10], (c) Coplanar crossover implemented by QCA cells in clock zones 1 and 3 in [11], and (d) Coplanar crossover implemented by QCA cells in clock zones 2 and 4 in [11].

zones 2 and 4) [11]. Cells in the switch phase can cross cells in the release phase and cells in the hold phase can cross cells in the relax phase. Signals can be transmitted normally because of no interaction between cells.

7.2 PROPOSED 5-INPUT MAJORITY GATE AND D FLIP-FLOP

7.2.1 PROPOSED 5-INPUT MAJORITY GATE

Figure 7.5 shows the proposed 5-input majority gate. The schematic of the 5-input majority gate is given in Figure 7.5(a). As shown in Figure 7.5(a), A, B, C, D, and E represent input cells, and F represents the output cell. Figure 7.5(b) shows the layout of the proposed 5-input majority gate. It contains 12 QCA cells, six of which are internal cells. Moreover, all input and output cells are not trapped by other cells, which can overcome the shortcoming of the previous designs in [12, 13]. The 5-input majority gate can be converted into a 3-input AND gate and an OR gate. To implement the AND and OR gates, as shown in Figure 7.5(c)-(d), two of the five inputs can be set to -1 or 1, respectively.

Figure 7.6 shows the simulation result of the proposed 5-input majority gate using QCA Designer [22], which indicates that the proposed gate can achieve the expected highly polarized output at clock 0. The values of A, B, C, D, and E are 1.00 and the output is 9.50e-001. This means that the output loses 5.00% of the input signal, which is within the allowable noise limit. As a result, the proposed 5-input majority gate maintains signal integrity. However, relying solely on simulation tools to verify the behavior of QCA circuits has shortcomings. Another method for the verification is to further verify the

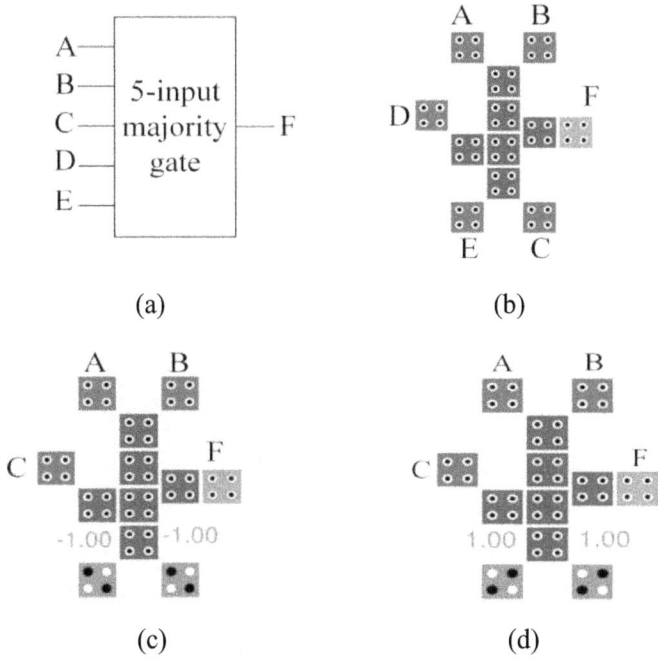

FIGURE 7.5 The proposed 5-input majority gate. (a) Schematic of the proposed 5-input majority gate, (b) Layout of the proposed 5-input majority gate, (c) Layout of the proposed 3-input AND gate based on (b), and (d) Layout of the proposed 3-input OR gate based on (b).

function of the circuit through physical calculations. The physical calculations involve calculating the Coulomb force of the electrons between the cells and then deducing the output cell's polarization value. The proposed 5-input majority gate is based on the response of QCA cells, and the output cell can be affected by all other cells to acquire accurate results. The following basic formulas are used in the calculation, where k is a constant, $q1$ and $q2$ are the Coulomb force, and r is the distance between electrons.

$$U = \frac{kq_1q_2}{r} \tag{3}$$

$$kq_1q_2 = 23.04 \times 10^{-9} \tag{4}$$

$$U_{T(SUM)} = \sum_{i=1}^{2} U_i \tag{5}$$

Figure 7.7 shows the cell positions of the proposed 5-input majority gate. In order to calculate the kink energy of any cell, the 5-input majority gate and the electrons in the cells are numbered. The electrostatic energy of different electrons in every cell are calculated when the input (A, B, C, D, E) = (1, 1, 1, 1, 1). Note that only two types

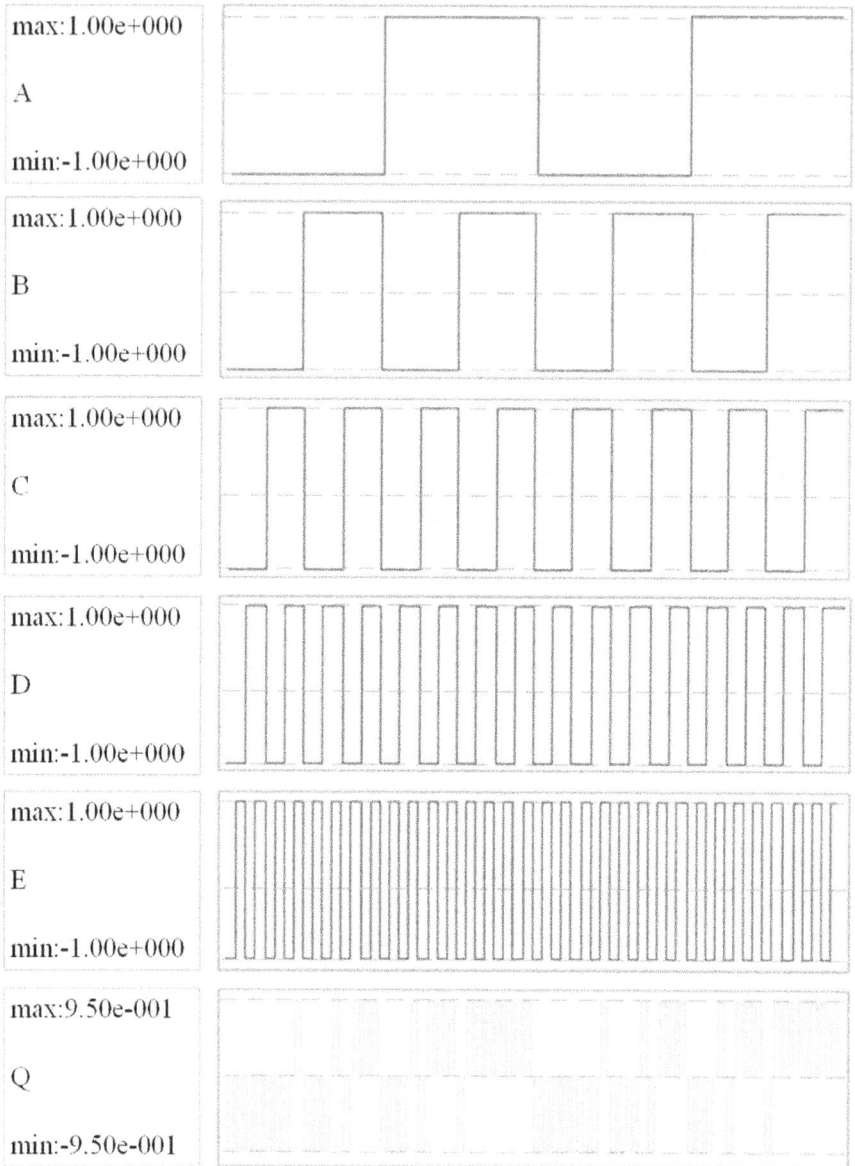

FIGURE 7.6 Simulation result for the proposed 5-input majority gate.

of outcomes (i.e. 1 and 0) are possible. Table 7.1 shows the calculated physical verification results by using Eq. (3)–(5). Eq. (3)–(4) can be used to calculate the electrostatic energy U_{Tx} or U_{Ty} between two different electrons. The total electrostatic energy UT(SUM) can be obtained from Eq. (5). As shown in Table 7.1, the electrostatic energy in Figure 7.7(a) is lower than that in Figure 7.7(b), indicating that the output

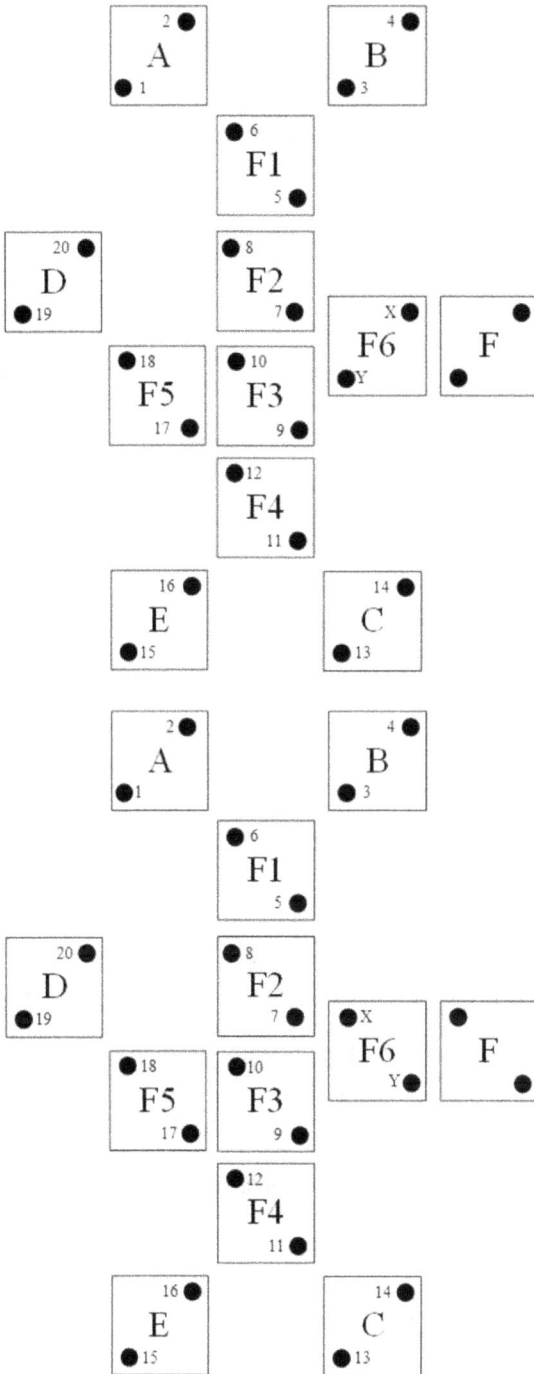

FIGURE 7.7   Cell positions of the proposed 5-input majority gate.

TABLE 7.1
Physical Verification Results

Case	Case 1 (when F = 1)	Case 2 (when F = 0)
U_{TX} (electron x)	14.55×10^{-20} (J)	10.11×10^{-20} (J)
U_{TY} (electron y)	9.63×10^{-20} (J)	15.11×10^{-20} (J)
$U_{T\,(SUM)}$	24.18×10^{-20} (J)	25.22×10^{-20} (J)

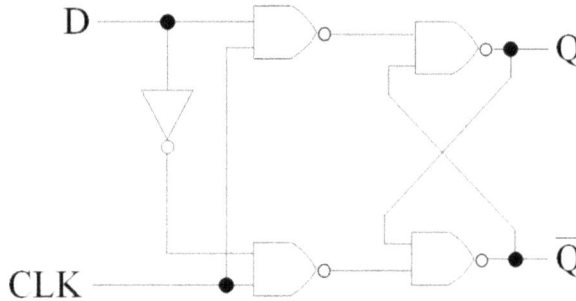

FIGURE 7.8   Circuit schematic of the traditional D flip-flop.

cell in Figure 7.7(a) is more stable than that in Figure 7.7(b). This also means that, all the cells are in the correct orientation. Since the analysis methods for other different types of inputs (A, B, C, D, E) are analogical with input (A, B, C, D, E) = (1, 1, 1, 1, 1), all other types of inputs can be analyzed in the same way.

7.2.2   PROPOSED D FLIP-FLOP

Figure 7.8 shows the circuit schematic of the traditional D flip-flop, which has cross lines. In a QCA circuit, the existence of cross lines will not only bring additional energy consumption to the circuit but also lead to a complex circuit layout. Therefore, the traditional D flip-flop is not suitable for the QCA circuit.

Figure 7.9 shows the D flip-flop that can be reconstructed based on a 5-input majority gate. As shown in Figure 7.9(a), the reconstructed D flip-flop is only composed of a 5-input majority gate. D and its reverse are used as two inputs of the majority gate. The clock signal, an input polarization value P = −1, and the output of the D flip-flop are the other three inputs. It can be seen from Figure 7.9(b) that the reconstructed D flip-flop implemented based on the 5-input majority gate only needs 15 cells. Figure 7.10 shows the simulation results of the proposed D flip-flop. It can be seen that the proposed D flip-flop's logic function is correct.

7.2.3   COMPARISONS

To make a fair and comprehensive comparison with the state-of-the-art 5-input majority gates (i.e. gates in [12–20]), the same simulation conditions described in the

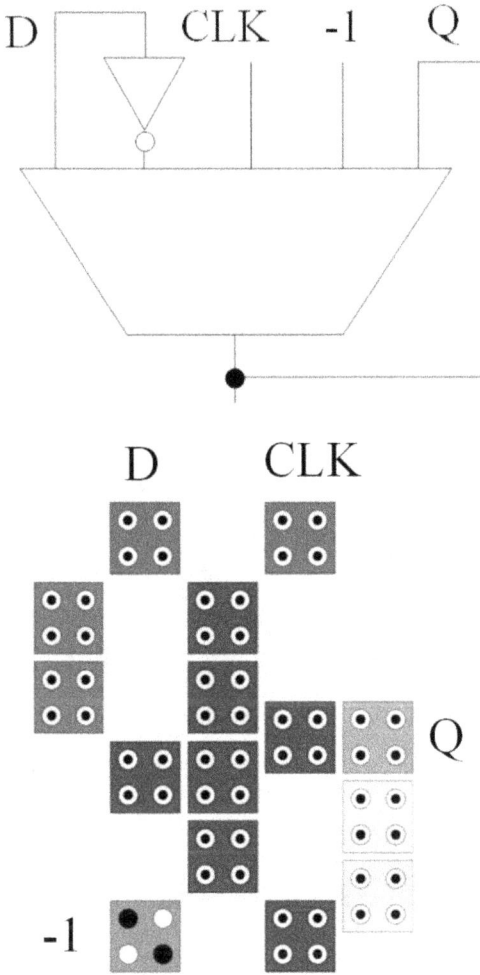

FIGURE 7.9 The proposed D flip-flop. (a) Schematic of the proposed D flip-flop, and (b) the proposed 5-input majority gate's layout.

above sections were used and the simulation tool was fixed to "bistable approxima-tion" with simulation parameters as shown in Table 7.2. In order to estimate the total and average energy dissipations of the proposed QCA-based structures, the QCA Designer-E tool was used [23].

A comparison between the proposed 5-input majority gate and the previous majority gate designs is shown in Table 7.3 considering different performance parameters. Figure 7.11 shows the histogram of the comparisons of 5-input majority gates. It can be seen that the proposed 5-input majority gate consumes less energy. After calculation, it can be found that the total energy dissipation and average energy dissipation per cycle of the proposed gate can be reduced by 37.42% and 35.70% at the cost of moderate cell count and area.

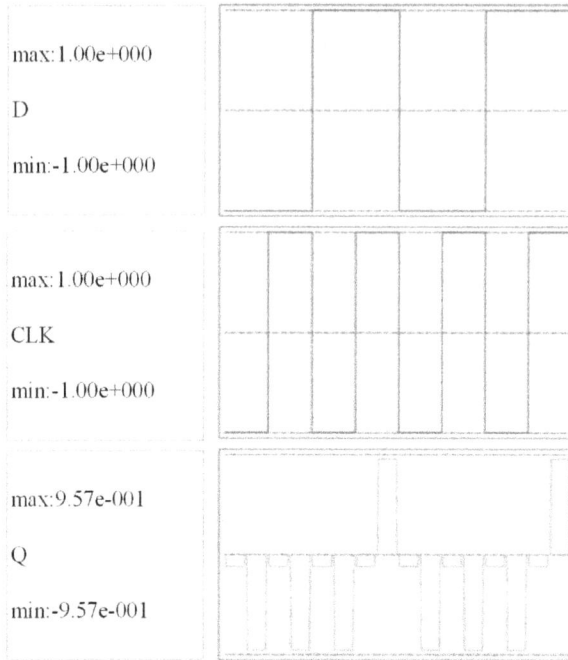

FIGURE 7.10 Simulation result for the proposed D flip-flop.

TABLE 7.2
Simulation Parameters of the QCA Designer Tool

Parameters	Values
Numbers of Samples	12.8×10^3
Convergence Tolerance	0.01×10^{-1}
Radius of Effect (nm)	65.00
Relative Permittivity	12.90
Clock High	9.80×10^{-22}
Clock Low	3.80×10^{-23}
Clock Shift	0.00
Clock Amplitude Factor	2.00
Layer Separation	11.50
Maximum Iterations per Sample	1.00×10^2
Cell Width (nm)	18.00
Cell Height (nm)	2.00
Dot Diameter (nm)	2.00

TABLE 7.3
Overhead Comparison Results Among the 5-Input Majority Gates

Design	Cell Count	Area (nm²)	Area utilization	Delay	Polarization	Total energy dissipation	Average energy dissipation
[12]	10	6.32	51.28	0.25	9.96	9.70	8.82
[13]	10	9.43	34.32	0.25	9.54	7.90	7.19
[14]	13	11.86	35.52	0.25	8.24	1.62	1.47
[15]	11	7.88	45.23	0.25	9.08	1.16	1.05
[16]	11	11.86	30.05	0.25	5.12	1.25	1.14
[17]	14	14.28	31.76	0.25	9.51	1.95	1.77
[18]	10	9.9	32.73	0.25	9.55	4.77	4.33
[19]	11	11.86	30.05	0.5	9.50	5.35	4.86
[20]	14	16.64	27.26	0.25	9.50	1.93	1.76
Proposed	12	13.82	28.13	0.25	9.50	1.47	1.30

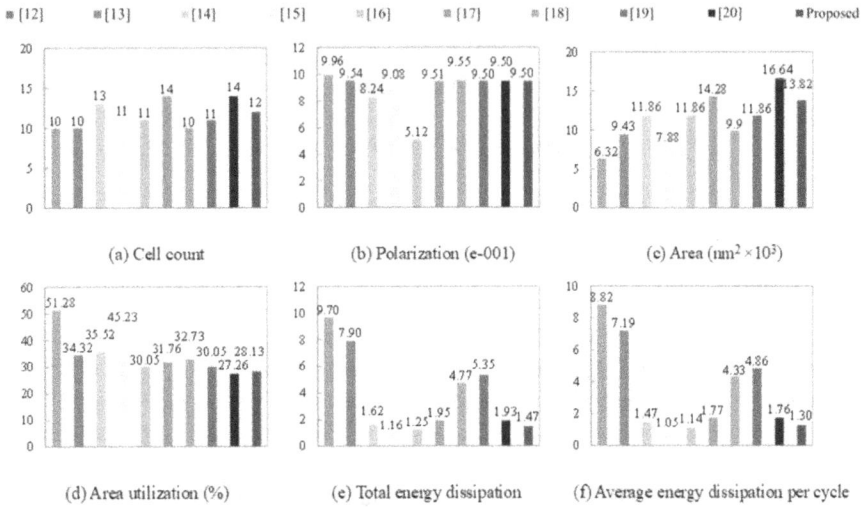

FIGURE 7.11 Histogram of the comparisons of 5-input majority gates.

In order to make a further detailed comparison with the other D flip-flops in terms of the area, delay, and cell count for the proposed D flip-flop, different overhead is calculated with Eq. (6, 7) [24]. Table 7.4 shows the overhead comparison results among the proposed D flip-flop and the previous designs. It can be seen from Table 7.3 that, compared with the existing D flip-flops, the proposed D flip-flop reduces the cell count, area, and delay by 35.24%, 27.78%, and 27.78%, respectively. The proposed D flip-flop consumes significantly less energy because it only uses a single 5-input

TABLE 7.4
Overhead Comparison Results Among The Proposed D Flip-Flop and the Previous Designs

D flip-flop	[19]	[20]	[21]	Proposed
Cell count	16	37	25	15
Area (um²)	0.02	0.04	0.03	0.02
Delay	0.50	0.75	1.00	0.50
Total energy dissipation	11.00	17.00	8.33	2.84
Average energy dissipation per cycle	10.00	15.40	7.57	2.58
ADP	0.01	0.03	0.03	0.01
ACP	0.16	1.11	0.75	0.15

majority gate. The total energy consumption is reduced by 74.46%, and the energy consumption per clock cycle is reduced by 74.46% on average, compared to alternative solutions. Moreover, the ADP and ACP are improved by 44.44% and 57.58%, respectively. In summary, the proposed D flip-flop has a lower overhead and can be used to design high-performance QCA sequential circuits and systems for quantum computing.

$$ADP = Area \times Delay \tag{6}$$

$$ACP = ADP \times Cell\,count \tag{7}$$

7.3 PROPOSED XOR GATE AND POLAR ENCODER

7.3.1 Proposed XOR Gate

In this section, a novel XOR/XNOR gate is proposed, which is based on the explicit Coulomb force between QCA cells to obtain correct outputs.

Figure 7.12 shows the proposed XOR/XNOR gate and Figure 7.13 shows the simulation results of design using the QCA Designer. The parameters in QCA Designer are default and the simulation engine is coherence vector engine. Table 7.5 shows the default parameters for the coherence vector simulation engine in QCA Designer. As shown in Figure 7.12(a-b), the proposed XOR/XNOR gate consists of 11 QCA cells with an area of 9204 nm² and a latency of 0.50 clock cycles. When the two enable inputs are set to (−1, 1), the XOR operation is performed. When the two enable inputs are set to (1, −1), the XNOR operation is performed. Figure 7.12(c-d) shows the two I/O access methods for the proposed design.

The simulation results of XOR gate are shown in Figure 7.13(a). When the two inputs {A, B} are set to {0, 0}, the output Q is '0'; when the {A, B} are set to {0, 1}, the output Q is '1'; when the {A, B} are set to {1, 0}, the output Q is '1'; when the {A, B} are set to {1, 1}, the output Q is '0'. The simulation results of XNOR gate

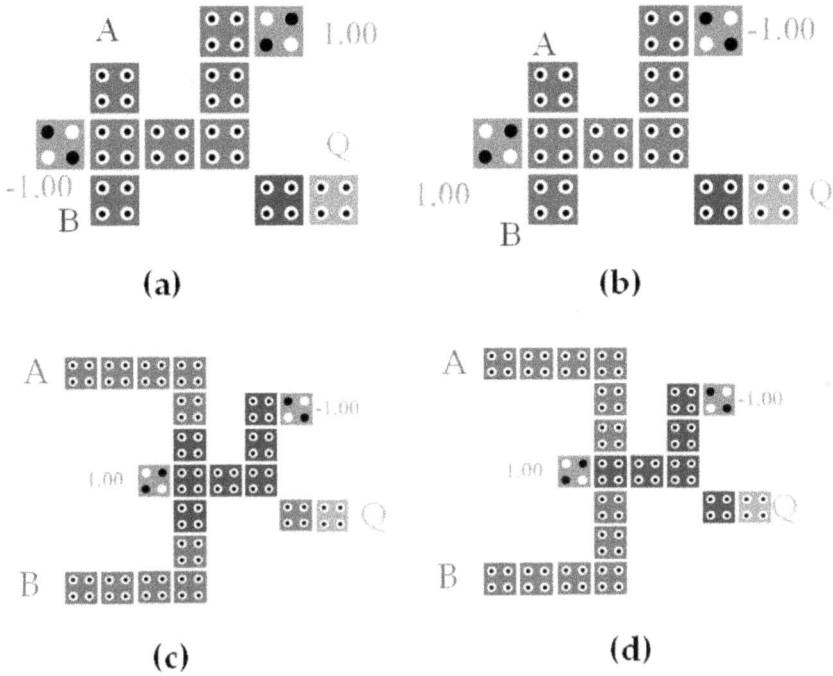

FIGURE 7.12 Proposed XOR/XNOR gate. (a) XOR gate, (b) XNOR gate, (c) the I/O access method and (d) the I/O access method.

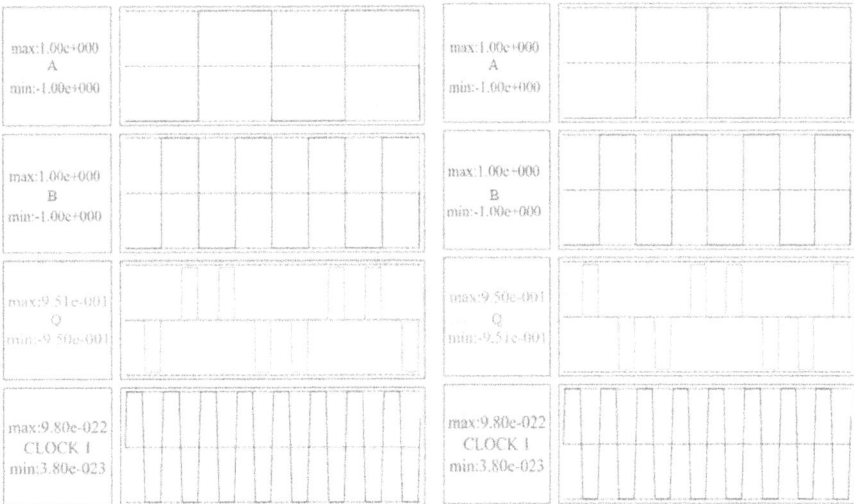

FIGURE 7.13 Simulation results of the proposed XOR/XNOR gate. (a) XOR gate and(b) XNOR gate.

TABLE 7.5
The Default Parameters for the Coherence Vector
Simulation Engine In QCA Designer

Simulation Engine	Coherence Vector
Temperature	1.00 K
Relaxation Time	1.00×10^{-15} s
Time Step	1.00×10^{-16} s
Cell size	18 nm × 18 nm
Gap of cell	2 nm
Radius of Effect	80 nm
Clock High	98×10^{-22} J
Clock Low	3.8×10^{-23} J
Clock Shift	0
Total Simulation Time	7.00×10^{-11} s
Relative Permittivity	12.9
Layer Separation	11.5 nm

are shown in Figure 7.13(b). When the {A, B} are set to {0, 0}, the output Q is '1';
when the {A, B} are set to {0, 1}, the output Q is '0'; when the {A, B} are set to {1,
0}, the output Q is '0'; when the {A, B} are set to {1, 1}, the output Q is '1'. It can
be seen from Figure 7.13 that the functioning of the XOR/XNOR gate is correct. The
waveform value of the XOR gate is in the range of (−9. 51× 10^{-1}~9. 50× 10^{-1}), and
the waveform value of the XNOR gate is in the range of (−9.50 × 10^{-1}~9.51 × 10^{-1}).

7.3.2   POLAR ENCODERS

In secure communications, encoders and decoders can be implemented by using QCA
circuits. Polar code was initially introduced by E. Arikan in 2009 [25]. Polar code is
a forward error correction code used for signal transmission. Polar code has explicit
proof for the channel performance which has now almost closed the gap to Shannon's
limit and is included as code for the control channels in the 5G standard.

Figure 7.14 shows the design of polar encoders in the QCA technology proposed
by Das et al [34]. If the inputs $x1$, $x2$, $x3$, and $x5$ in Figure 7.14 are set to '0', it can
act as an (8, 4) polar encoder (G(8, 4)). Formulas (4)–(11) are Boolean expressions for
the eight outputs of G(8, 4).

$$y1 = x1 \oplus x2 \oplus x3 \oplus x4 \oplus x5 \oplus x6 \oplus x7 \oplus x8 \tag{8}$$

$$y2 = y5 \oplus y6 \oplus y7 \oplus y8 \tag{9}$$

$$y3 = x3 \oplus x4 \oplus x7 \oplus x8 \tag{10}$$

$$y4 = y7 \oplus y8 \tag{11}$$

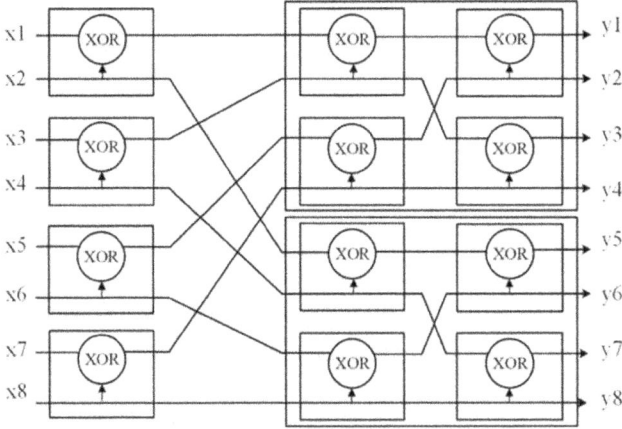

FIGURE 7.14 Schematic diagram of a polar encoder [27].

$$y5 = x2 \oplus x4 \oplus x6 \oplus x8 \tag{12}$$

$$y6 = x6 \oplus x8 \tag{13}$$

$$y7 = x4 \oplus x8 \tag{14}$$

$$y8 = x8 \tag{15}$$

We propose and have implemented the coplanar G(8, 4) circuit using the proposed XOR gate. Figure 7.15 shows the layout of the proposed G(8, 4) design. Figure 7.16 shows the simulation results of the G(8, 4) design in QCA Designer. The simulation engine of QCA Designer was set to the coherence vector engine, the total simulation time was set to 7.000000e-010 s, and other parameters were set to the default values. The proposed G(8, 4) consists of 518 cells. Its area is 0.635 μm² and its latency is 3.50 clock cycles. The total and average energy dissipation of the proposed G(8, 4) are 1.08×10^{-1} eV and 9.84×10^{-3} eV.

7.3.3 COMPARISONS

Table 7.6 shows the comparisons of the circuit performance and energy dissipation of the existing XOR/XNOR gates. The area was calculated through layouts, the latency was calculated by identifying the number of used clock cycles, and the total and average energy dissipation of these gates were estimated with the QCA Designer-E tool. Note that the simulation parameters of the QCA Designer-E were set as default values.

It can be seen from Table 7.6 that the cell count and area of the proposed gate are larger than those of the designs in [31,33]. However, the total and average energy dissipation of the designs in [31,33] are larger than those of the proposed XOR/XNOR

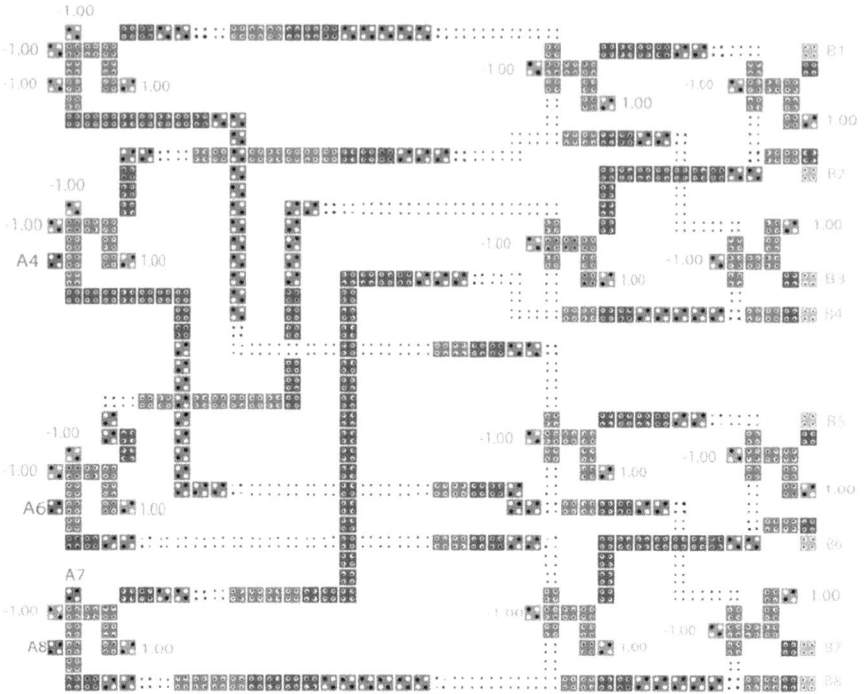

FIGURE 7.15   The QCA implementation of the proposed G(8, 4) circuit.

gate. It can be seen from Table 7.6 that the energy dissipation of the proposed XOR/XNOR gate is smaller than the designs in [26–28,30–33]. Therefore, the proposed gate is competitive, especially in terms of energy dissipation. The proposed XOR/XNOR gate can be used in large QCA circuits with low power.

Table 7.7 shows the comparisons of the performance and energy consumption of the proposed G(8, 4) circuit. The cell count, area, and latency of the proposed G(8, 4) circuit is lower compared with alternative solutions and the proposed G(8, 4) has the lowest energy dissipation. Compared with the best solution in [35], the cell count and area of the proposed G(8, 4) are reduced by 13.67% and 12.05%, respectively, and the latency is reduced by 0.25 clock cycles. Therefore, the proposed (8, 4) polar encoder is competitive in terms of cell count, area, and energy consumption.

7.4   CONCLUSIONS

This chapter has first proposed a 5-input majority gate and it can be used to implement highly efficient single-layer QCA circuits. The proposed majority gate has many advantages, including single-layer implementation, lower energy consumption, and moderate complexity and area. Second, the proposed majority gate has been used to effectively design a D flip-flop and the proposed flip-flop has low cell count and low energy dissipation. Thirdly, an XOR/XNOR gate has been proposed

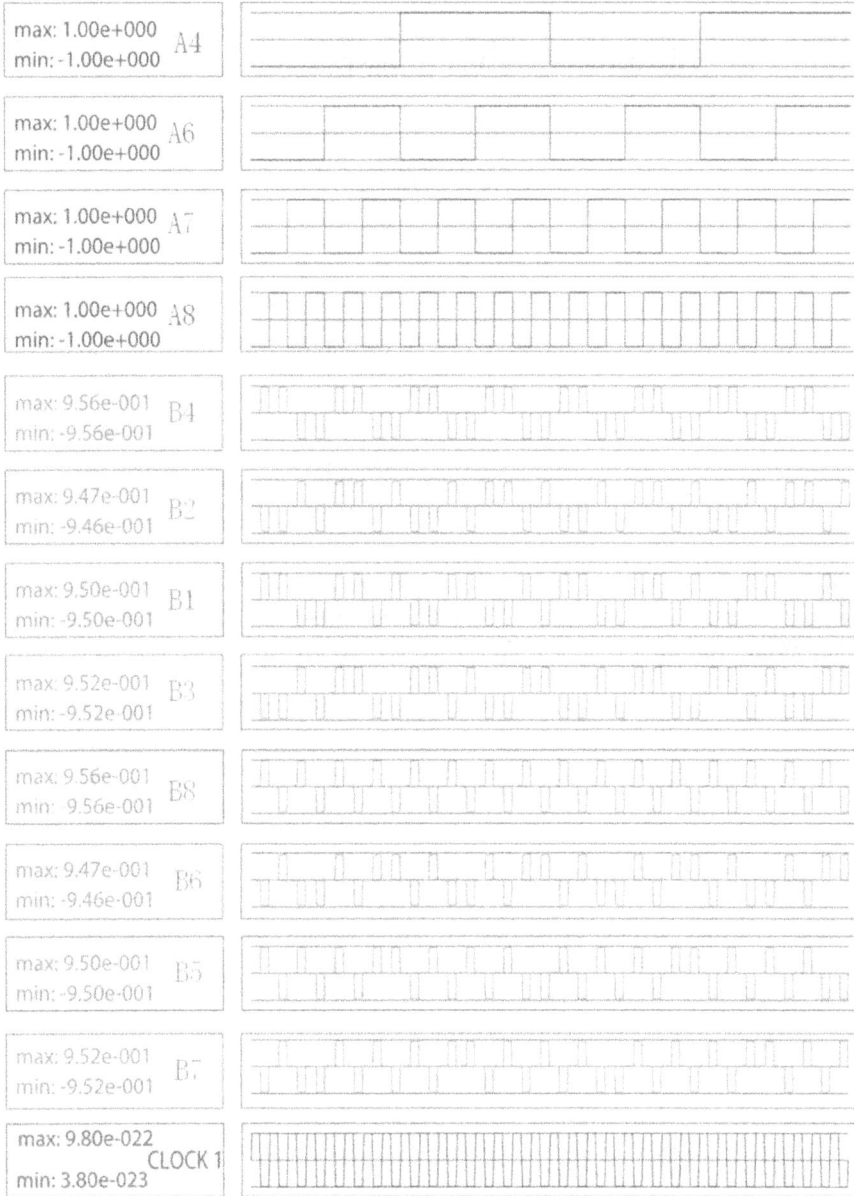

FIGURE 7.16 Simulation Results of the Proposed G(8, 4) Circuit.

and implemented. A polar encoder is designed based on the proposed XOR/XNOR gate. Compared with the state-of-the-art G(8, 4), the proposed G(8, 4) has a 13.67% reduction in terms of cell count and a 12.05% reduction in terms of area, also with low energy dissipation. The input signals of the proposed G(8, 4) circuits can output the signals faster due to the reduction in latency. Simulations using QCA Designer

TABLE 7.6
Comparisons of the Performance And Energy Dissipation For Alternative Xor/Xnor Gates

Design	Cell Count	Area (nm²)	Latency	Total Energy Dissipation (eV)	Average Dissipation (eV)
[26]	14	11,564	0.25	1.17×10^{-2}	1.07×10^{-3}
[27]	13	11,564	0.50	8.06×10^{-3}	7.33×10^{-4}
[28]	14	16,284	0.50	1.44×10^{-2}	1.31×10^{-3}
[29]	9	7644	0.25	3.56×10^{-3}	3.24×10^{-4}
[30]	13	13,524	0.50	8.02×10^{-3}	7.29×10^{-4}
[31]	10	5684	0.50	9.78×10^{-3}	8.89×10^{-4}
[32]	12	9604	0.25	$1,07 \times 10^{-2}$	9.76×10^{-4}
[33]	10	6084	0.50	1.13×10^{-2}	1.02×10^{-3}
Pro-XOR/XNOR	11	9204	0.50	4.13×10^{-3}	3.75×10^{-4}

TABLE 7.7
Comparisons of the Performance And Energy Dissipation For Alternative (8, 4) Polar Encoders

Design	Cell Count	Area (μm²)	Latency	Total Energy Dissipation (eV)	Average Energy Dissipation (eV)
[34]	1188	1.915	6.25	3.29×10^{-1}	2.99×10^{-2}
[35]	600	0.722	3.75	1.13×10^{-1}	1.02×10^{-2}
Proposed	518	0.635	3.50	1.08×10^{-1}	9.84×10^{-3}

and QCA Designer-E confirm that the proposed designs have outperformed all prior designs and yielded significant improvements in almost all criteria. In the future, these design schemes can be considered to be applied to the design of QCA circuits.

REFERENCES

[1] J. Fortes, "Future Challenges in VLSI System Design," *IEEE Computer Society Annual Symposium on VLSI*, pp. 5–7, 2003.

[2] C. Lent, P. Tougaw, W. Porod, et al. "Quantum Cellular Automata," *Nanotechnology*, vol. 4, no. 1, pp. 49–57, 1993.

[3] P. Tougaw, C. Lent, "Logical Devices Implemented Using Quantum Cellular Automata," *Journal of Applied Physics*, vol. 75, no. 3, pp. 1818–1825, 1994.

[4] C. Lent, P. Tougaw, "A Device Architecture for Computing with Quantum Dots," *Proceedings of the IEEE*, vol. 85, no. 4, pp. 541–557, 1997.

[5] G. Toth, C. Lent, "Quantum Computing with Quantum-dot Cellular Automata," *Physics Review*, vol. A63, no. 2000, 1–9, 2000.

[6] M. Parish, "Modeling of Physical Constraints on Bistable Magnetic Quantum Cellular Automata," Ph.D. Thesis, University of London, UK, 2003.

[7] E. Blair, C. Lent, "Quantum-dot Cellular Automata: An Architecture for Molecular Computing," *International Conference on Simulation of Semiconductor Processes and Devices*, pp. 14–18, 2003.

[8] G. Tóth, C. Lent, "Quasi-adiabatic Switching for Metal-island Quantum-dot Cellular Automata," *Journal of Applied Physics*, vol. 85, pp. 2977–2984, 1999.

[9] C. Lent, M. Liu, Y. Lu, "Bennett Clocking of Quantum-dot Cellular Automata and the Limits to Binary Logic Scaling," *Nanotechnology*, vol. 17, no. 16, pp. 4240–4251, 2006.

[10] A. Gin, P. Tougaw, S. Williams, "An Alternative Geometry for Quantum-dot Cellular Automata," *Journal of Applied Physics*, vol. 55, no. 12, pp. 8281–8286, 1999.

[11] S. Shin, J. Jeon, K. Yoo, "Wire-crossing Technique on Quantum-dot Cellular Automata," *NGCIT2013, the 2nd International Conference on Next Generation Computer and Information Technology*, pp. 52–57, 2013.

[12] K. Navi, S. Sayedsalehi, R. Farazkish, et al, "Five-input Majority gate, A New Device for Quantum-dot Cellular Automata," *Journal of Computational and Theoretical Nanoscience*, vol. 7, pp. 1546–1553. 2010.

[13] K. Navi, R. Farazkish, S. Sayedsalehi, et al, "A New Quantum-dot Cellular Automata Full-adder," *Microelectronics Journal*, vol. 41, pp. 820–826. 2010.

[14] A. Roohi, H. Khademolhosseini, S. Sayedsalehi, et al, "A Symmetric Quantum-dot Cellular Automata Design for 5-input Majority gate," *Journal of Computational Electronics*, vol. 13, no. 1, pp. 701–708, 2014.

[15] T. N. Sasamal, A. K. Singh, A. Mohan, "An Optimal Design of Full Adder Based on 5-input Majority Gate in Coplanar Quantum-dot Cellular Automata," *Optik*, vol. 127, pp. 8576–8591, 2016.

[16] T. Sasamal, A. Singh, A. Mohan, "An Efficient Design of Quantum-dot Cellular Automata Based 5-input Majority Gate with Power Analysis," *Microprocessors and Microsystems*, vol. 59, pp. 103–117, 2018.

[17] M. B. Khosroshahy, M. H. Moaiyeri, K. Navi, et al, "An Energy and Cost Efficient Majority-based Ram Cell in Quantum-dot Cellular Automata," *Results in Physics*, vol. 7, pp. 3543–3551, 2017.

[18] R. Jaiswal, T. N. Sasamal, "Efficient Design of Exclusive-or Gate Using 5-input Majority Gate in Qca," *IOP Conference*, pp. 1–6, 2017.

[19] F. Zheng, G. Xie, X. Wang, "Analysis and Design of Coplanar Five-Input Mjority Gate on Quantum-Dot Cellular Automata," *Acta Electronica Sinica*, vol. 48, no. 5, pp. 0372–2112, 2020.

[20] T. Sasamal, A. Mohan, A. Singh, "Optimal Realization of Full Adder in QCA Using 5-input Majority Gate," *2020 International Conference on Industry 4.0 Technology*, 2020.

[21] R. S. Kumaresan, M. Raj, L. Gopalakrishnan, "Area-Efficient D-Flip Flop and XOR in QCA," *International Conference on Computing, Communication and Networking Technologies*, pp. 1–5, 2020.

[22] K. Walus, G. Jullien, "Design Tools for an Emerging SoC Technology: Quantum-dot Cellular Automata," *Proceedings of the IEEE*, vol. 94, no. 6, pp. 1225–1244, Jun. 2006.

[23] F. S. Torres, R. Wille, P. Niemann, et al, "An Energy-aware Model for the Logic Synthesis of Quantum-dot Cellular Automata," *IEEE Transactions on Computer-Aided Design of Integrated Circuits and Systems*, vol. 37, no. 12, pp. 3031–3041, 2018.

[24] B. Sen, M. Goswami, S. Some, et al, "Design of Sequential Circuits in Multilayer QCA Structure," *International Symposium on Electronic System Design*, pp. 21–25, 2014.

[25] E. Arikan, "Channel Polarization: A Method for Constructing Capacity-achieving Codes for Symmetric Binary-input Memoryless Channels," *IEEE Transactions on Information Theory*, vol. 55, no. 7, pp. 3051–3073, 2009.

[26] A. Chabi, A. Roohi, R. DeMara, et al, "Cost-efficient QCA Reversible Combinational Circuits Based on a New Reversible Gate," *2015 18th CSI International Symposium on Computer Architecture and Digital Systems (CADS)*, pp. 1–6, 2015.

[27] M. Berarzadeh, S. Mohammadyan, K. Navi, et al, "A Novel Low Power Exclusive-OR via Cell Level-based Design Function in Quantum Cellular Automata," *Journal of Computational Electronics*, vol. 16, no. 3, pp. 875–882, 2017.

[28] F. Ahmad, G. Bhat, H. Khademolhosseini, et al, "Towards Single Layer Quantum-dot Cellular Automata Adders Based on Explicit Interaction of Cells," *Journal of Computational Science*, vol. 16, no. 1, pp. 8–15, 2016.

[29] H. Chen, H. Lv, Z. Zhang, et al, "Design and Analysis of a Novel Low-power Exclusive-OR Gate Based on Quantum-dot Cellular Automata," *Journal of Circuits, Systems and Computers*, vol. 28, no. 8, pp. 1–18, 2019.

[30] L. Xingjun, S. Zhiwei, C. Hongping, et al, "A New Design of QCA-based Nanoscale Multiplexer and Its Usage in Communications," *International Journal of Communication Systems*, vol. 33, no. 4, pp. 1–14, 2020.

[31] R. Laajimi, M. Niu, "Nanoarchitecture of Quantum-Dot Cellular Automata (QCA) Using Small Area for Digital Circuits," *Advanced Electronics Circuits–Principles, Architectures and Applications on Emerging Technologies*, vol. 1, no. 1, pp. 67–84, 2018.

[32] A. Majeed, M. Zainal, E. Alkaldy, et al, "Single-bit Comparator in Quantum-dot Cellular Automata (QCA) Technology Using Novel QCA-XNOR Gates," *Journal of Electronic Science and Technology*, vol. 19, no.3, pp. 1–12, 2021.

[33] B. Safaiezadeh, E. Mahdipour, M. Haghparast, et al, "Design and Simulation of Efficient Combinational Circuits Based on a New XOR Structure in QCA Technology," *Optical and Quantum Electronics*, vol. 53, no. 12, pp. 1–16, 2021.

[34] J. Das, D. De, "QCA Based Design of Polar Encoder Circuit for Nano Communication Network," *Nano Communication Networks*, vol. 18, no. 1, pp. 82–92, 2018.

[35] S. Ahmed, S. Naz, S. Bhat, "Design of Quantum-dot Cellular Automata Technology Based Cost-efficient Polar Encoder for Nanocommunication Systems," *International Journal of Communication Systems*, vol. 33, no. 18, pp. 1–14, 2020.

8 Physically Realizable Reversible Logic Gates in Beyond CMOS QCA Technology

Lakshminarayanan Gopalakrishnan,
Marshal R, Raja Sekar K, Seok-Bum Ko
and Anantharaj Thalaimalai Vanaraj

8.1 INTRODUCTION

Columbic repulsion based computation and communication makes QCA an interesting choice to realize high speed circuits at smaller nanoscale levels. Information transmission and computation using the position of electrons and repulsion between them requires very low power and they are an exciting alternative to design ultra low power circuits [1, 2]. Several researches are done to find molecules and metals for QCA realization in room temperature.

Reversible logic is another interesting paradigm that can be used to design highly energy efficient circuits. Reversible gates have the ability to regenerate the inputs from the outputs [3]. The inherent pipelined and clock controlled information transmission in QCA makes them an ideal technology to design reversible gates. The direction of information transmission can be modified by controlling the clock signal applied to the cells. However, the challenges involved with the fabrication of QCA circuits increases the complexity and difficulty in making a physically realizable QCA circuits [4–6]. To overcome the limitations and challenges of conventional clocking scheme based design in QCA, where clocking is more congested, novel clocking schemes are proposed in the literature.

Although novel clocking schemes are proposed with different cell zone arrangement, the amount of circuits designed using such clocking schemes is very less. The reversible gates are less explored in QCA. In order to exploit the advantages of novel clocking schemes and reversible logic, reversible logic gates are proposed using USE clocking scheme. The proposed gates can be extended and used to design reversible circuits. The proposed reversible gates include Feynman, Toffoli, Peres, and Fredkin gate. The proposed blocks can be combined to realize any combinational and sequential circuit in QCA. The array based cell zone reduces the complexity of the fabrication of QCA circuits. Hence, the proposed designs are more physically realizable compared to other reversible QCA gate designs.

DOI: 10.1201/9781003361633-8

The major contributions in this chapter include:

- design of Feynman gate using USE clocking scheme
- design of Toffoli gate using USE clocking scheme
- design of Peres gate using USE clocking scheme
- design of Fredkin gate using USE clocking scheme
- validation of proposed designs in QCADesigner.

The rest of the chapter is organized in the following structure. Section 2 discusses about the elements of QCA in brief and the conventional clocking used in QCA. Section 3 focuses on the novel clocking schemes available to make physically realizable QCA circuits. Section 4 provides a brief overview on reversible logic and reversible gates. Section 5 presents the proposed reversible gates using USE clocking scheme and compares the proposed designs with other designs. Section 6 provides the conclusions to the chapter.

8.2   CLOCKING IN QCA

QCA operation is dependent upon its cells and majority logic. The cells have quantum dots and electrons. Cells with different combinations of quantum dots and electrons are available. The predominantly used combination is four quantum dots and two electrons. Cells are defined as normal and rotated cells based on the position of dots. Columbic repulsion inside the cell and with the adjacent cells determines the position of electrons in each cell. As there are only two possible orientation of electrons possible inside such cells, the positions are encoded with binary values for computation and communication using the electrons. As the position of electrons in each cell is dependent upon the maximum orientation positions of the adjacent cells, majority logic determines the computation and communication of information along with columbic repulsion [7, 8].

The cells have four phases of operation. In order to have a controlled information transfer and computation, the cells are connected to clocks and the phases of each cell are controlled to ensure that the information is flowing in a particular direction and race conditions are avoided by ensuring that the information flows in a pipelined manner. The clock signal is divided in to four clocks and they are named as clock 0 to clock 3. Each clock has a $\pi/2$ phase difference with adjacent zone clocks. Depending upon the state of the clock, the cell will be in one of its operation state. Cells are grouped and connected to clocks depending upon communication direction. As no clock will be in an identical state to another clock, the cells in different clock zones will also be in different states. The clock signal is switched accordingly so that the information flows in a particular direction [9]. Figure 8.1(a) shows wires formed using QCA cells. Figure 8.1(b) shows the state of the cell during each phase of the clock. Figure 8.1(c) shows how the cells will be operating without overlapping of states at any point of time and the phase difference between clocks. Figure 8.1(d) shows the information flow can be controlled by connecting the cells to different clock zones to achieve both directed flow and pipelined operation.

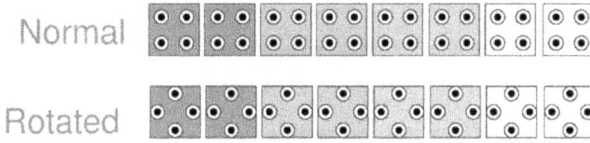

FIGURE 8.1(a) Wires using QCA cells.

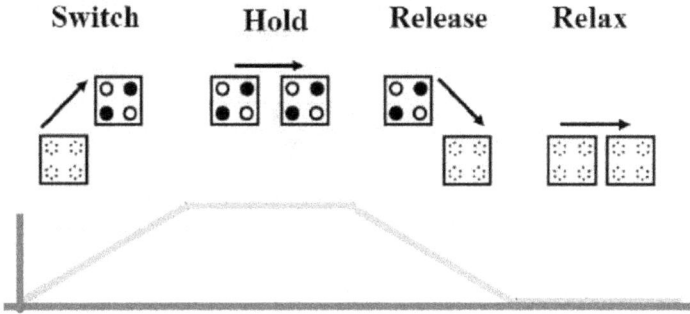

FIGURE 8.1(b) Operational states of QCA cell.

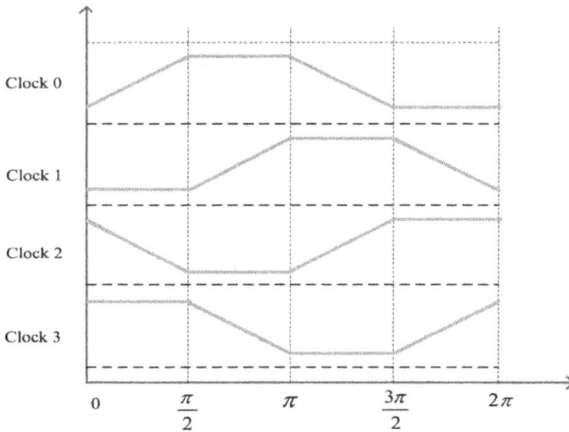

FIGURE 8.1(c) QCA clocks and their phase differences.

8.3 CLOCKING SCHEMES IN QCA

The conventional clocking scheme discussed in the previous section is effective. However, the small size of QCA cells increases the complexity of such circuits. The extremely small nanoscale structure of QCA cells makes them more vulnerable to fabrication defects such as omission and misplacement defects. In addition to that,

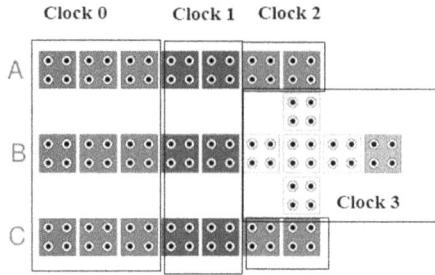

FIGURE 8.1(d)   Controlled information flow using clock zones.

FIGURE 8.2   1-D clocking scheme.

placing cells connected to different clock zones in closer proximity increases the complexity of the clocking. It makes fabrication more complex and increases the probability of fabrication defects. In order to reduce the fabrication complexity due to the clocking mechanism, novel mechanisms are proposed by researchers by arranging cells in clock zones in an array manner. This ensures that there is sufficient spacing available for wire crossing of clock signals in QCA circuits.

One-Dimensional (1-D) clocking scheme was initially proposed in [10] to solve the complexity issues arising out of wire crossings of clock signals. The major limitation with 1-D clocking scheme in [10] was the length of the wire and the difficulty in providing feedback path. Combinational circuits are developed in the literature using the 1-D scheme [11–13]. In [13],1-D spatial clocking scheme was proposed. It reduced the issue of long wires. However, the major limitation and challenge faced by the clocking scheme was the difficulty in feedback paths. It increases the difficulty in developing sequential circuits using the 1-D clocking scheme. Figure 8.2 shows the different 1-D clocking schemes.

In [14], a two dimensional clocking scheme was proposed by dividing each column by small clock zones and the wire length was reduced. However, the scheme still faced challenges with designing sequential circuits. The 2-DDW wave clocking scheme in [14] had zones arranged in diagonal manner to reduce the wire crossing complexity. It supported bi-directional input and output to increase the throughput. The 2-DDW wave technique offered reduced complexity, increased throughput, well defined clocking circuitry, and reduced thermodynamic effects. However, the complexity of designing sequential circuits was still high as the inputs have to flow through multiple zones for providing a feedback. Figure 8.3 shows the two dimensional and 2-DDW wave clocking schemes.

In [15], a USE clocking scheme was proposed. The USE scheme inherited the advantages of systolic array based arrangement in 2DDW clocking scheme and used multiple adjacent clock zone patterns to provide the feedback paths. The USE

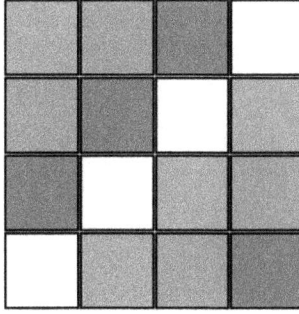

FIGURE 8.3   2-DDW clocking scheme.

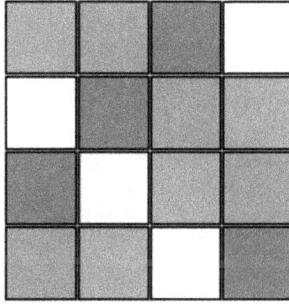

FIGURE 8.4   USE clocking scheme.

scheme reduced the feedback path delay by a large extent compared to 2DDW clocking scheme. The electric field generation circuitry was proposed with metal wires placed diagonally under the circuit layer. The USE scheme supports multilayer wire crossings which increases the complexity of fabrication. However, multilayer circuits have reduced delays compared to coplanar crossings done by using normal and rotated cells. Figure 8.4 shows the USE clocking scheme.

In [16], a Robust, Efficient, and Scalable (RES) clocking scheme is introduced. It has the features of USE scheme, but it supports coplanar wire crossing. However, the RES scheme has increased clocking wire crossing complexity compared to the USE scheme. In [17], an optimized 2-D clocking scheme is introduced to reduce the wire crossing complexity. Instead of having a square shaped clock zone, a rectangular clock zone was used with a ratio of 2:1. However, the shift in the consecutive rows increases the fabrication complexity. In [17], an Efficient, Scalable, and Regular (ESR) clocking scheme is introduced by incorporating the advantages of USE, RES and by reducing the wire crossings. However, the clock zone array size considered is smaller and it increases the fabrication complexity. Figure 8.5 shows the RES, optimized 2-D and ESR clocking schemes.

The major concern among all the clocking schemes is the reduction in metal wire crossings for fabrication. The clock zones also need to be larger to ensure that the

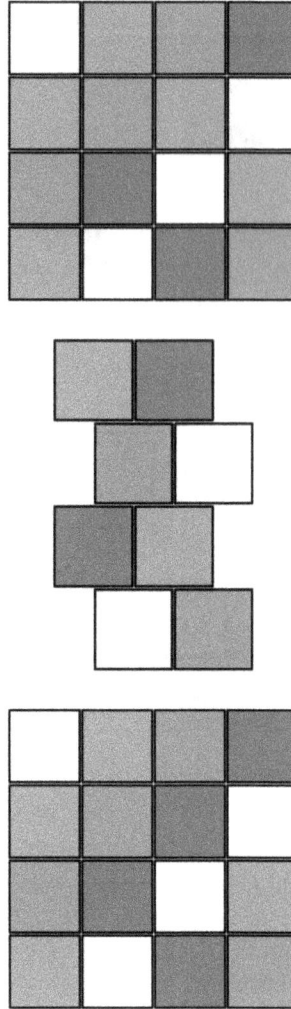

FIGURE 8.5   Clocking schemes. (a) RES, (b) Optimized 2-D, (c) ESR clocking scheme.

fabrication complexity is minimal. A trade-off needs to be achieved between the size of the clock zones, wire crossings and delay paths especially with the feedback delays.

8.4   REVERSIBLE LOGIC AND REVERSIBLE GATES

The computations performed by the existing devices are irreversible. The inputs cannot be identified from the outputs alone. According to Launder's principle [18], the irreversible operations dissipate $KTln2$ Joules of energy, where K represents the Boltzmann's constant, T represents the temperature. For a room temperature operation at 300 degree Kelvin, the amount of energy dissipated will be around 2.8×10^{-21} J. The dissipation is quite minimal, however it has to be noted that it cannot be considered negligible in ultra low power devices[19]. In 1973, Bennet showed that a system can dissipate no energy if they are able to return to their initial state irrespective of the changes that occurred in it [20]. Reversible logic circuits help to return the systems to their initial states irrespective of the changes occurring in the system. The reversible circuits are also called lossless circuits as they neither loose information nor energy.

Reversible gates have a 1 to 1 mapping between the input and output. This facilitates the input reconstruction from the output. Hence, a reversible logic gate has 'm' inputs and 'm' outputs. Some reversible gates have constant inputs also called as ancillary inputs for maintaining the reversible functionality. Reversible logic gates may also have additional outputs that are generated in addition to the desired functionality. Such outputs are known as Garbage outputs. They have no significance in generating the desired output. However, they play a key role in generating the input back from the output to maintain the reversibility functionality. For an efficient reversible circuit, the count of reversible logic gates, ancillary inputs and garbage outputs must be as low as possible. These parameters determine the cost of the circuit. The widely used reversible gates are Peres, Toffoli, Feynman, and Fredkin gates [21–31].

8.4.1 FEYNMAN GATE

The Feynman gate is a 2×2 gate. One output is an exclusive-Or output of two inputs. Another Output is same as the input. The functionality of Feynman gate is presented in Table 8.1. Figure 8.6 shows the logical representation of Feynman gate. The logical expressions of Feynman gate are given by (1) and (2).

$$P = A \tag{1}$$

$$Q = A \oplus B \tag{2}$$

TABLE 8.1
Feynman gate logical operation

A	B	P	Q
L	L	L	L
L	H	L	H
H	L	H	H
H	H	H	L

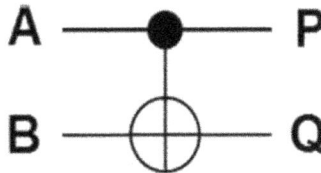

FIGURE 8.6   Feynman gate.

8.4.2 TOFFOLI GATE

The Toffoli gate is a 3×3 gate. The first two outputs are unchanged values of two inputs. The last output will be inverted value of third input if the remaining inputs are High. Figure 8.7 shows the representation of Toffoli gate and Table 8.2 provides the logical operation functionality of Toffoli gate. The logical expressions for the outputs of Toffoli gate are given by (3)–(5).

$$P = A \tag{3}$$

$$Q = B \tag{4}$$

$$R = AB \oplus C \tag{5}$$

8.4.3 FREDKIN GATE

The Fredkin gate is a 3×3 gate. Two outputs are having identical value as the inputs. The third input determines the selection of input that will be reflected by the input. Figure 8.8 shows the representation of Fredkin gate and Table 8.3 provides the logical operation of Fredkin gate. The logical expressions for Fredkin gate are given (6)–(8).

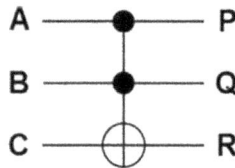

FIGURE 8.7   Toffoli gate.

TABLE 8.2
Toffoli gate logical operation

A	B	C	P	Q	R
L	L	L	L	L	L
L	L	H	L	L	H
L	H	L	L	H	L
L	H	H	L	H	H
H	L	L	H	L	L
H	0	H	H	0	H
H	H	0	H	H	H
H	H	H	H	H	0

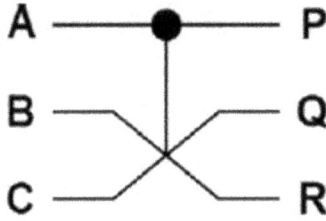

FIGURE 8.8 Fredkin gate.

TABLE 8.3
Fredkin gate logical operation

A	B	C	P	Q	R
L	L	L	L	L	L
L	L	H	L	L	H
L	H	L	L	H	L
L	H	H	L	H	H
H	L	L	H	L	L
H	L	H	H	H	L
H	H	L	H	L	H
H	H	H	H	H	H

$$P = A \tag{6}$$

$$Q = \overline{A}B + AC \tag{7}$$

$$R = AB + \overline{A}C \tag{8}$$

8.4.4 PERES GATE

Peres gate is a 3×3 gate. One of the outputs reflects the input. One output is an exclusive-or output of two inputs. The last output is a combination of AND and exclusive-or operation. Figure 8.9 shows the representation of Peres gate and Table 8.4 provides the logical operation of Peres gate. The logical expressions of Peres gate are given by (9)–(11).

$$P = A \tag{9}$$

$$Q = A \oplus B \tag{10}$$

$$R = AB \oplus C \tag{11}$$

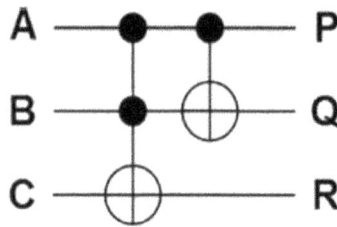

FIGURE 8.9 Peres gate.

TABLE 8.4
Peres gate logical operation

A	B	C	P	Q	R
L	L	L	L	L	L
L	L	H	L	L	H
L	H	L	L	H	L
L	H	H	L	H	H
H	L	L	H	H	L
H	L	H	H	H	H
H	H	L	H	L	H
H	H	H	H	L	L

8.5   REVERSIBLE LOGIC QCA GATES

Among the clocking schemes discussed in Section 3, the USE scheme is an efficient scheme as it has limited disadvantages compared to other schemes. As reversible logic gates are not realized using the novel clocking schemes, reversible logic gates using USE scheme is proposed in this section to showcase the realization of reversible gates using novel clocking schemes. The proposed designs have better physical realization possibilities as the clock zones are regularly arranged. Figure 8.10 shows the proposed Feynman, Toffoli, Fredkin, and Peres gate using USE clocking scheme and multilayer crossover. The results of simulation are presented in Figure 8.11. Table 8.5 compares the proposed designs with other reversible logic gate designs in QCA. The area is computed by using the area. The proposed gates have more cell count, area, delay, and ADP compared to conventional clock-based designs realized using conventional clocking scheme. This is mainly due to the size of clock zones and the arrangement of clock zones. In conventional clock scheme, feedback can be provided easily. But in the novel clocking schemes, feedback can be provided only through the predefined clock zone arrangement of the clocking scheme.

8.6   CONCLUSION

QCA is an interesting circuit design nanotechnology that can be utilized to realize high performance systems. Reversible logic is another promising technique to design ultra

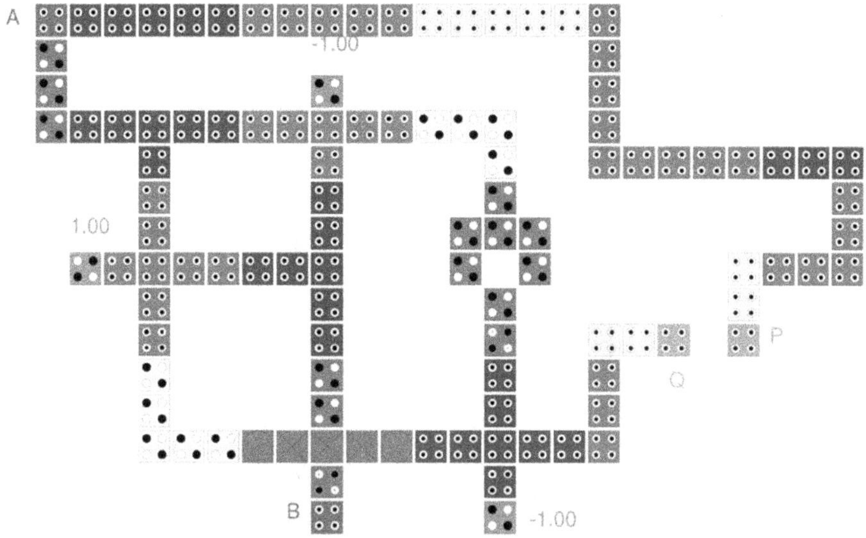

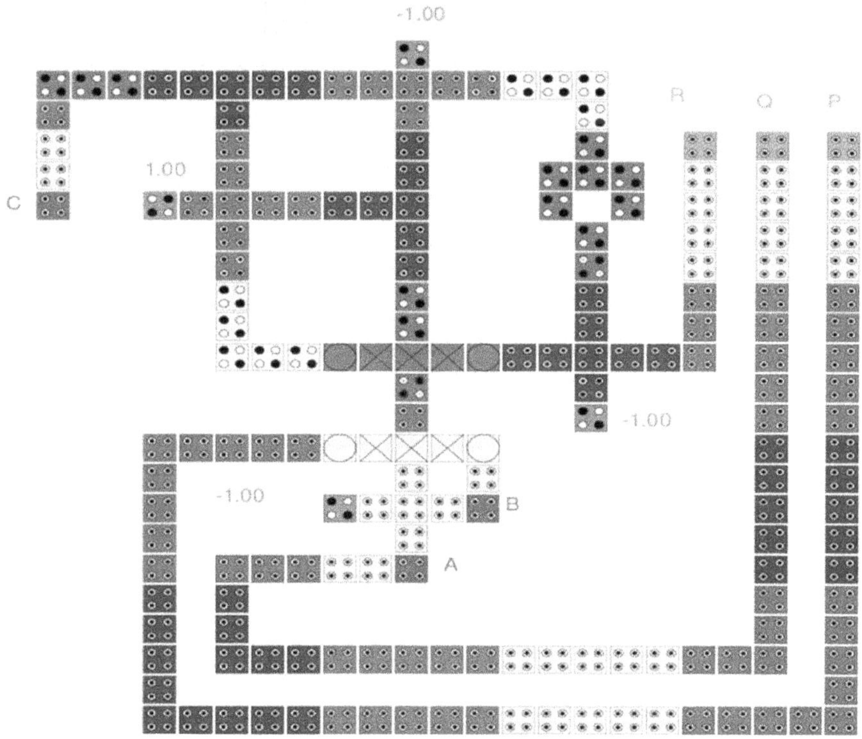

FIGURE 8.10   Proposed circuits. (a) Feynman, (b) Toffoli, (c) Fredkin, (d) Peres.

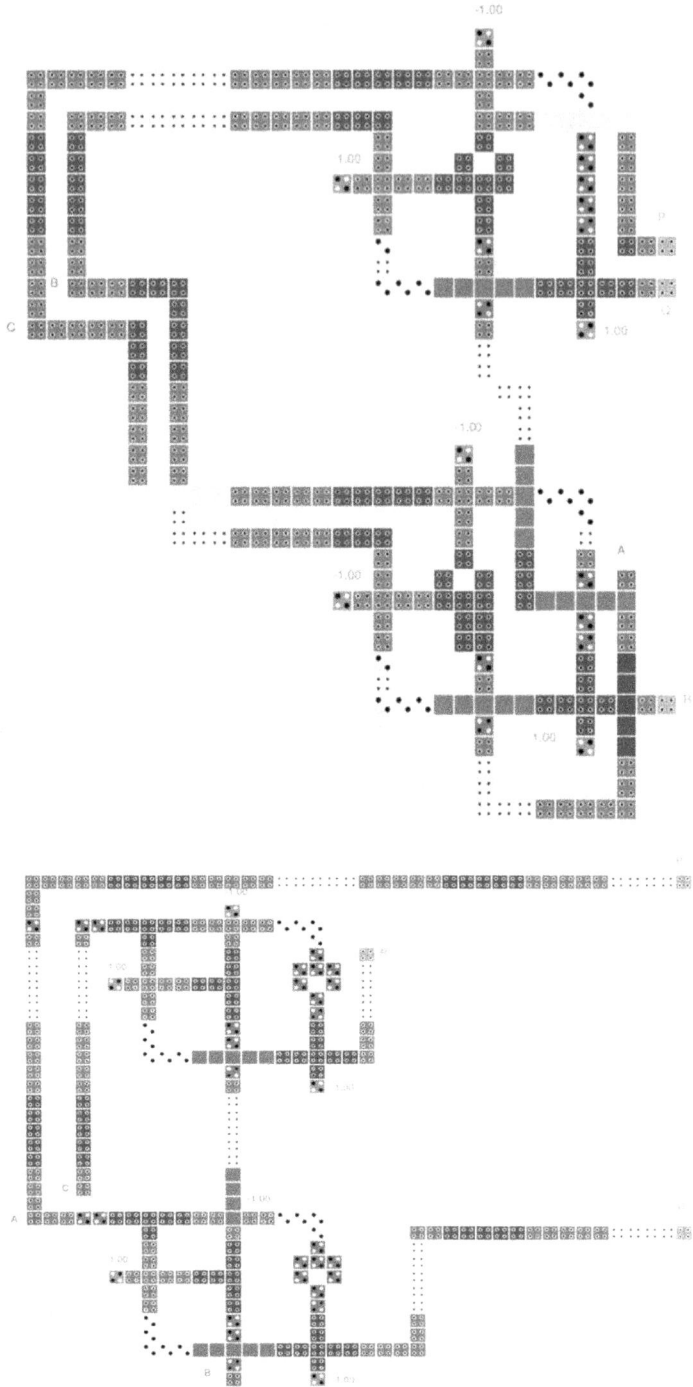

FIGURE 8.10 (Continued)

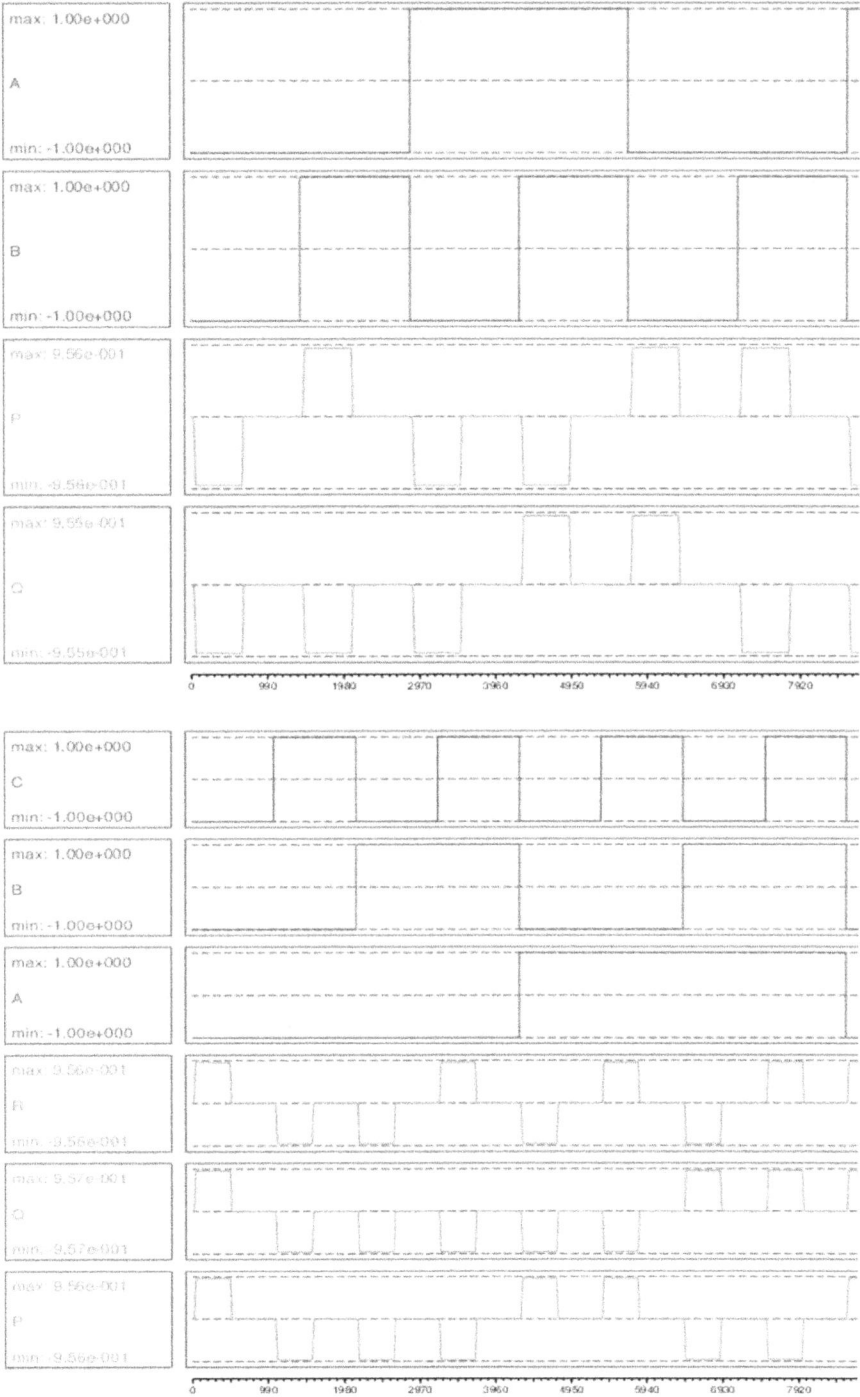

FIGURE 8.11 Proposed circuits validation. (a) Feynman, (b) Toffoli, (c) Fredkin, (d) Peres.

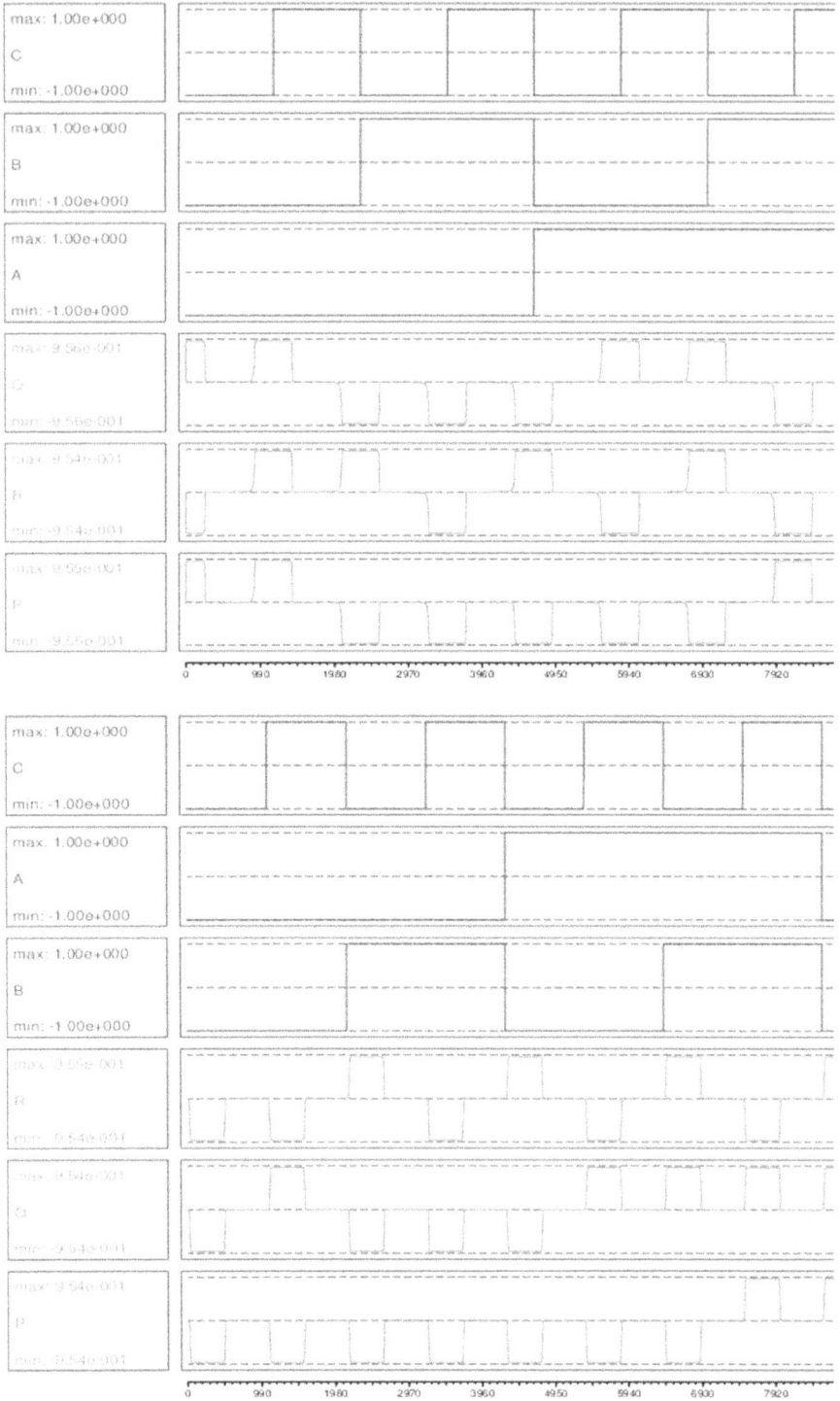

FIGURE 8.11   (Continued)

TABLE 8.5
Proposed gates comparison

Ref.	Cell Count	Area in (μm²)	Delay in Clock phases	Area-Delay Product (ADP)	Clocking Scheme
Feynman gate					
[21]	43	0.04	3	0.12	Conventional
[22]	54	0.04	3	0.12	Conventional
[23]	32	0.03	3	0.09	Conventional
[31]	11	0.01	2	0.02	Conventional
Proposed	114	0.14	8	1.12	USE
Toffoli gate					
[24]	170	0.21	4	0.84	Conventional
[25]	128	0.20	4	0.80	Conventional
[23]	45	0.04	3	0.12	Conventional
[31]	19	0.01	2	0.02	Conventional
Proposed	319	0.52	11	5.72	USE
Fredkin gate					
[26]	246	0.37	4	1.48	Conventional
[27]	102	0.08	3	0.24	Conventional
[23]	73	0.06	3	0.18	Conventional
[31]	42	0.04	3	0.12	Conventional
Proposed	195	0.23	9	2.07	USE
Peres Gate					
[28]	99	0.10	4	0.40	Conventional
[29]	137	0.13	4	0.52	Conventional
[30]	97	0.08	4	0.32	Conventional
[31]	29	0.02	2	0.04	Conventional
Proposed	273	0.57	12	6.84	USE

low power circuits. The clock controlled information transfer mechanism and inherent pipelining features in QCA make them an ideal technology to design reversible logic circuits. Reversible logic gates are already available using conventional clocking scheme in QCA. Conventional clocking scheme based QCA circuits have higher probability of fabrication defects and realization complexity. Novel clocking schemes are available in literature to develop physically realizable QCA circuits. In this chapter, reversible logic gates are realized using USE clocking scheme. The proposed designs are scalable and can be used to design higher complex reversible logic based systems.

REFERENCES

1. Huang, J., Momenzadeh, M., and Lombardi, F. (2007). An overview of nanoscale devices and circuits. *IEEE Design and Test of Computers*, 24(4), 304–311. https://doi.org/10.1109/mdt.2007.121
2. Lent, C. S., Tougaw, P. D., Porod, W., and Bernstein, G. H. (1993). Quantum cellular automata. *Nanotechnology*, 4(1), 49–57. https://doi.org/10.1088/0957-4484/4/1/004.

3. Shende, V. V., Prasad, A. K., Markov, I. L., and Hayes, J. P. (2003). Synthesis of reversible logic circuits. *IEEE Transactions on Computer-Aided Design of Integrated Circuits and Systems*, 22(6), 710–722. https://doi.org/10.1109/tcad.2003.811448.

4. Yang, X., Cai, L., Wang, S., Wang, Z., and Feng, C. (2012). Reliability and performance evaluation of QCA devices with rotation cell defect. *IEEE Transactions on Nanotechnology*, 11(5), 1009–1018. https://doi.org/10.1109/tnano.2012.2211613

5. Chetti, S. C., and Yatgal, O. (2022). QCA: A Survey and Design of Logic Circuits. *Global Transitions Proceedings*. https://doi.org/10.1016/j.gltp.2022.04.012.

6. Sefidi, M. S., Abedi, D., and Moradian, M. (2013). Design a collector with more reliability against defects during manufacturing in nanometer technology, QCA. *Journal of Software Engineering and Applications*, 6(6), 304–312. https://doi.org/10.4236/jsea.2013.66038

7. Snider, G. L., Orlov, A. O., Amlani, I., Bernstein, G. H., Lent, C. S., Merz, J. L., and Porod, W. (1999). Quantum-dot cellular automata: Line and majority logic gate. *Japanese Journal of Applied Physics*, 38(12S), 7227. https://doi.org/10.1143/jjap.38.7227

8. Tougaw, P. D., and Lent, C.S. (1994). Logical devices implemented using quantum cellular automata. *Journal of Applied Physics*, 75(3), 1818–1825, 1994. https://doi.org/10.1063/1.356375

9. Sasamal, T. N., Singh, A. K., and Mohan, A. (2020). Clocking Schemes for QCA. In: Trailokya Nath Sasamal, Ashutosh Kumar Singh, and Anand Mohan (eds) Quantum-Dot Cellular Automata Based Digital Logic Circuits: A Design Perspective. Studies in Computational Intelligence, vol 879. Springer, Singapore, pp. 139–145. https://doi.org/10.1007/978-981-15-1823-2_9

10. Lent, C. S., and Tougaw, P. D. (1997). A device architecture forcomputing with quantum dots. *Proceedings of the IEEE*, 85(4), 541–557. https://doi.org/10.1109/5.573740

11. Hennessy, K., and Lent, C. S. (2001). Clocking of molecularquantum-dot cellular automata. *Journal of Vacuum Science & Technology B: Microelectronics and Nanometer Structures Processing, Measurement, and Phenomena*, 19(5), 1752–1755. https://doi.org/10.1116/1.1394729

12. Janez, M., Pecar, P., and Mraz, M. (2012). Layout design ofmanufacturable quantum-dot cellular automata. *Microelectronics Journal*, 43(7), 501–513. https://doi.org/10.1016/j.mejo.2012.03.007

13. Rani, S., and Sasamal, T. N., (2017). Design of QCA circuits using new 1D clocking scheme. International Conference on Telecommunication and Networks, 1–6. https://doi.org/10.1109/TEL-NET.2017.8343540

14. Vankamamidi, V., Ottavi, M., and Lombardi, F. (2007). Two-dimensional schemes for clocking/timing of QCAcircuits. *IEEE Transactions on Computer-Aided Design of Integrated Circuitsand Systems*, 27(1), 34–44. https://doi.org/10.1109/TCAD.2007.907020

15. Campos, C. A. T., Marciano, A. L., Neto, O. P. V., and Torres, F. S. (2016). USE:A Universal, Scalable and Efficient clocking scheme for QCA. *IEEE Transactions on Computer-Aided Design of Integrated Circuits and Systems*, 35(3), 513–517. https://doi.org/10.1109/TCAD.2015.2471996

16. Goswami, M., Mondal, A., Mahalat, M. H., Sen, B., and Sikdar, B. K. (2019). An efficient clocking scheme for quantum-dot cellular automata. *International Journal of Electronics Letters*, 8(1), 1–14.https://doi.org/10.1080/21681724.2019.1570551

17. Pal, J., Pramanik, A. K., Sharma, J. S. *et al.* (2021). An efficient, scalable, regular clocking scheme based on quantum dot cellular automata. *Analog Integrated Circuits Signal Processing*, 107, 659–670. https://doi.org/10.1007/s10470-020-01760-4

18. Landauer, R. (1961). Dissipation and heat generation in the computing process. *IBM Journal of Research and Development*, 5, 183–191

19. Pan, W. D., and Nalasani, M. (2005) Reversible logic. *IEEE Potentials*, 24(1), 38–41. https://doi.org/10.1109/MP.2005.1405801

20. Bennett, C. H., (1973). Logical reversibility of computation. *IBM Journal of Research and Development*, 17, 525–532. https://doi.org/10.1147/rd.176.0525

21. Das, J. C., and De, D. (2016) Quantum-dot cellular automata based reversible low power parity generator andparity checker design for nanocommunication. *Frontiers of Information Technology & Electronic Engineering*, 17, 224–236. https://doi.org/10.1631/FITEE.1500079

22. Debnath, B., Das, J. C., and De, D. (2017) Reversible logic-based image steganography using quantumdotcellular automata for secure nanocommunication. *IET Circuits, Devices and Systems*, 11, 58–67. https://doi.org/10.1049/iet-cds.2015.0245

23. Singh, G., Sarin, R. K., and Raj, B. (2017) Design and analysis of area efficient QCA based reversible logicgates. *Microprocessors and Microsystems*, 52, 59–68. https://doi.org/10.1016/j.micpro.2017.05.017

24. Mohammadi, Z., and Mohammadi, M. (2014) Implementing a one-bit reversible full adder using quantum-dotcellular automata. *Quantum Information Processing*, 13, 2127–2147. https://doi.org/10.1007/s11128-014-0782-2

25. Garg, U., and Jain R., (2016) Design and performance analysis of reversible RSG gate using QCA. *International Journal of Computers and Applications*, 139, 37–41. https://doi.org/10.5120/ijca2016909506

26. Thapliyal, H., Ranganathan, N., and Kotiyal, S., (2013) Design of testable reversible sequential circuits. *IEEE Transactions on Very Large Scale Integration*, 21(7), 1201–1209. https://doi.org/10.1109/TVLSI.2012.2209688

27. Das, J. C., and De, D. (2017). Nanocommunication network design using QCA reversible crossbar switch. *Nano Communication Networks*, 13, 20–33. https://doi.org/10.1016/j.nancom.2017.06.003

28. Sarker, A., Bahar, A. N., Biswas, P. K., and Morshed, M. (2014). A novel presentation of Peres gate (PG) inquantum-dot cellular automata (QCA). *European Science Journal*, 10(21), 101–106. https://doi.org/10.19044/esj.2014.v10n21p%25p

29. Kumawat, R., Sasamal, T. N. (2016) Design of 1-bit and 4-bit adder using reversible logic in quantum-dotcellular automata. IEEE International Conference on Recent Trends in Electronics, Information & Communication Technology, 593–597. https://doi.org/10.1109/RTEICT.2016.7807891

30. Das, J. C., and De, D. (2016) Novel low power reversible binary incrementer design using quantum-dot cellular automata. *Microprocessors and Microsystems*, 42, 10–23. https://doi.org/10.1016/j.micpro.2015.12.004

31. Abutaleb, M. M. (2018). Robust and efficient QCA cell-based nanostructures of elementary reversible logic gates. *Journal of Supercomputing,* 74(11), 6258–6274. https://doi.org/10.1007/s11227-018-2550-z

9 Design of New Circuits for Reversible ALU in QCA Technology

Saeed Ghorbani Rizi[1], Abdalhossein Rezai[2] and Hamid Mahmoodian[3]*

[1]ACECR Institute of Higher Education, Isfahan Branch, Isfahan, Iran, saghorbani95@yahoo.com

[2]Department of Electrical Engineering, University of Science and Culture, Tehran, Iran, rezai@usc.ac.ir

[3]ACECR Institute of Higher Education, Isfahan Branch, Isfahan, Iran, mahmoodian@jdeihe.ac.ir

*Corresponding author

9.1 INTRODUCTION

Nowadays, the CMOS technology has limitations such as short channel effects in digital circuits design. An alternative to replace this technology is the Quantum-Dot Cellular Automata (QCA) technology [1, 2]. Several computational circuits such as shift register circuits [3, 4], full adder circuits [2, 5, 6, 7, 8], multiplexer circuits [1, 9, 10, 11], and Arithmetic and Logic Unit (ALU) circuits [12, 13, 14, 15, 16, 17, 18, 19, 20, 21, 22] have been implemented in this technology. In addition, Bennet [23] has demonstrated that the reversible logics are able to avoid energy loss. So, the design of reversible digital circuits became the focus of researchers' attention. The reversible circuits have an equal number of outputs and inputs. In addition, any input is mapped to an exclusive output, which reduces energy dissipation.

On the other hand, the main and basic section of any Central Processing Unit (CPU) is the ALU. This unit is able to implement operations such as OR, AND, XOR, subtractor, and full adder. So, several attempts [12, 13, 14, 15, 16, 17] have been done to enhance the efficiency of the Reversible ALU (RALU) circuits.

In this chapter, a Reversible Logic Unit (RLU) with low area and low latency is designed. Then, three 1-bit 3-layer QCA RALU circuits are developed with this RLU. To implement these RALU circuits, the QCADesigner tool is used. The results show that the RLU circuit is composed of 40 cells and $0.07~\mu m^{2~area}$. The first, second and third 1-bit QCA RALU circuits are composed of 209 ($0.29\mu m^2$), 174 ($0.24\mu m^2$), and 596 ($0.96\mu m^2$) cells (area), respectively. The results of the comparison show that our QCA RALU circuits have benefits compared to previous QCA RALU circuits regarding area, complexity, latency, and cost.

This chapter is set out as follows: Section 2 sets out the background in the QCA technology and an examination of related previous RALU circuits. In Section 3, the proposed QCA RLU is developed. Then, three new QCA RALU circuits are proposed. In Section 4, the designed circuits are simulated. The results of the simulation of RALU are reported and designed RALU have been compared with other RALU circuits. Finally, the conclusions are set out in Section 5.

9.2 BACKGROUND

9.2.1 QCA Cells

The quantum cells are the main building block and the smallest unit in the QCA technology. Every QCA cell consists of four dots and two moving electrons. Because of electrostatic repulsion between electrons, the electrons will be situated only at the angles of the QCA cell in the form of the diagonal [16, 24]. Therefore, there are two stable states in the QCA cell that are indicated in Figure 9.1.

9.2.2 QCA Wire

A QCA wire is composed of 90° or 45° cells in a row, which in the first case, the cells have equal polarizations and in the second case, the cells have different polarizations. Wire crossing are as follows in the QCA technology: crossing with different clock zones, multilayer crossing, and coplanar crossing [25].

The multilayer crossing uses at least three layers and two wires containing 90° cells cross in three layers. In coplanar crossing method, two wires 90° and 45° pass over

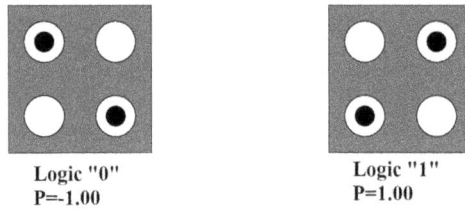

Logic "0"
P=-1.00

Logic "1"
P=1.00

FIGURE 9.1   Stable states for the cell in the QCA technology.

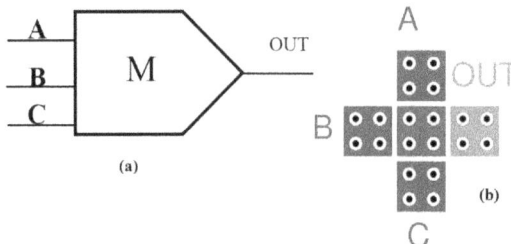

FIGURE 9.2   3-input majority gate (a) Logic, (b) QCA layout.

each other in one layer. The crossing with different clock zones uses one layer. In this method, two QCA wires can cross each other with a difference of two clock phases [26].

9.2.3   QCA GATES

Inverter and majority gates are basic QCA gates. Figure 9.2 displays a 3-input majority gate. The output of this gate is computed as follows [27].

$$OUT = M(C,B,A) = BC + BA + AC \tag{1}$$

The inverter gate output is the complement of its input, i.e. if the input is "0", the output is "1", else the output is "0". This gate is indicated in Figure 9.3.

9.2.4   QCA CLOCKING

Data in the QCA circuits is transferred from inputs to output using the clock. The QCA clock is composed of four phases: (1) switch, (2) hold, (3) release, and (4) relax that indicate in Figure 9.4.

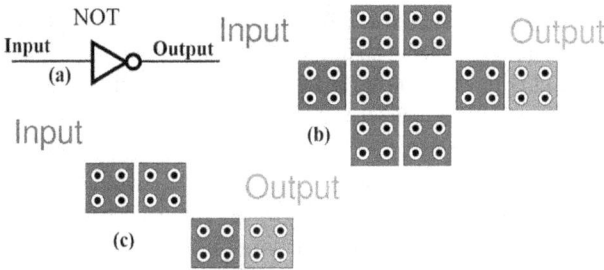

FIGURE 9.3   Inverter gates types (a) Logic circuit, (b) Robust inverter, (c) Basic inverter.

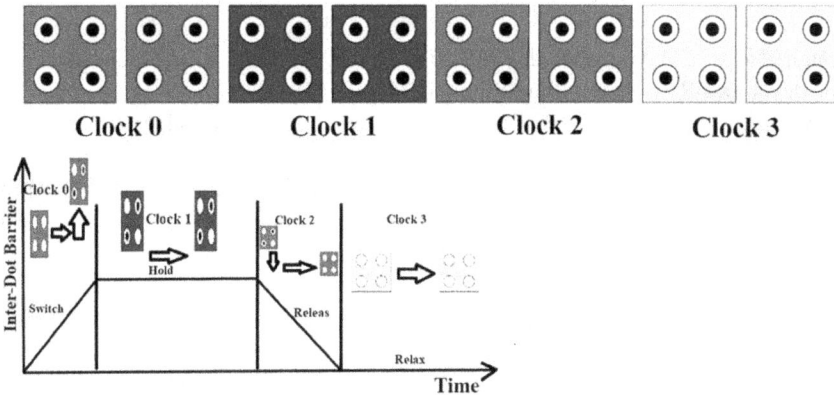

FIGURE 9.4   QCA clocking.

In the switching phase, the cell is unpolarized and the inter-dot potential barrier is low. In the holding phase, the potential barrier is high. In the release phase, the potential barriers begin to decrease and in the relaxing phase, the barriers hold down and cell is unpolarized [8].

9.2.5 RELATED WORKS

Chaves et al. [12] have reported a one-bit RALU circuit in the QCA technology. This circuit uses four Fredkin gates and two Toffoli gates. This circuit is designed in three layers that includes 1097 QCA cells. This circuit executes six logic operations.

Sen et al. [13] have presented a one-bit RALU circuit in the QCA technology. The presented ALU in [13] uses a reversible multiplexer. This circuit has two sections: the logic unit and the arithmetic unit, which are reversible. This circuit is designed in one layer that includes 2370 cells. It executes 16 operations.

Sasamal et al. [14] have provided a 1-bit RALU circuit using reversible multiplexer. This coplanar circuit can execute 16 arithmetic and logical operations.

Naghibzadeh and Houshmand [15] have presented a 1-bit RALU circuit in the QCA technology. This circuit presents a novel reversible gate called NHG. This RALU contains the NHG gate, Feynman gate, and Fredkin gate. This circuit is designed in three layers that includes 670 cells. This circuit executes 16 operations.

Oskouei and Ghaffari [16] have proposed a 1-bit RALU circuit. This circuit, which is designed in three layers, includes 332 cells. It executes four operations including AND, OR, XOR, and full adder. The designed RALU circuit in [16] has 0.38 μm^2 area.

Norouzi et al. [17] have proposed a 1-bit RALU. This circuit uses three Fredkin gates and an HNG gate. It can perform 20 operations. This RALU is designed in one layer using 480 cells.

Safaiezadeh et al. [18] have reported a 1-bit RALU circuit. This circuit uses a Fredkin gate, a BS1 gate, and two Feynman gates. The presented RALU can perform 20 operations. This RALU, which is designed in one layer, includes 350 cells. The developed RALU circuit in [18] has 0.51 μm^2 area.

Roy et al. [19] have proposed a one-bit RALU circuit. This circuit, which is designed in six layers, includes 230 cells. It can execute four operations. The developed RALU circuit in [19] has 0.1 μm^2 area.

Roy et al. [20] have proposed a one-bit RALU circuit. This design, which is designed in seven layers, includes 228 cells. This RALU can execute four operations. The developed RALU circuit in [20] has 0.14 μm^2 area.

Ahmadpour et al. [21] have proposed a 1-bit RALU circuit. This RALU is designed in one layer using 625 cells. The developed RALU circuit in [21] has 1.34 μm^2 area.

Rahimpour and Jafari [22] have proposed a 1-bit RALU circuit. This circuit, which is designed in three layers, includes 464 cells and it executes four operations. The developed RALU circuit in [22] has 0.78 μm^2 area.

9.3   THE PRESENTED RALU CIRCUITS

In this section, three 1-bit QCA RALU circuits are developed by using the reversible units. The presented RALU circuits carry out four operations: AND, NAND, XOR, and addition.

9.3.1 THE DEVELOPED RLU

The developed RLU is shown in Figure 9.5.

The developed RLU is designed in three layers, includes a Toffoli gate, a Feynman gate, and a NOT gate. To design the reversible logic, the Toffoli gate and Feynman gate, which are proposed in [28], are used. The inputs are C, B and A and the outputs are NAND (R'), XOR (R2), and P2, where C = 0 is constant input and P2 is garbage output. It's composed of 40 cells and occupies 0.07 $\mu m^{2 \, area}$.

9.3.2 THE FIRST PROPOSED RALU

The first presented RALU is indicated in Figure 9.6. It includes the proposed RLU unit, full adder and multiplexer gates presented in Figures 9.5, in [29], and in [10], respectively.

This designed RALU is implemented in three layers. Table 9.1 displays the 4:1 MUX operation table.

In this RALU, the inputs C, B, and A are designed in 1st layer. In addition, the 4:1 multiplexer is designed in such a way that the output, OUT, is shown in 1st layer. The 2nd layer is implemented as an interface to transfer information between the 1st layer and the 3rd layer. The full adder and the MUX are implemented in three layers. The

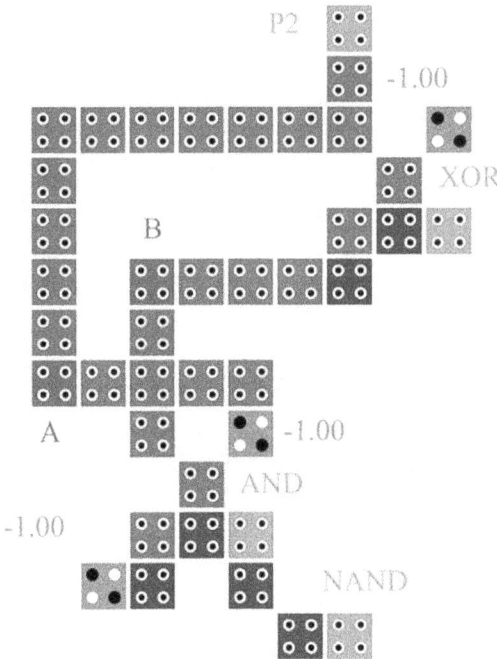

FIGURE 9.5   The developed RLU.

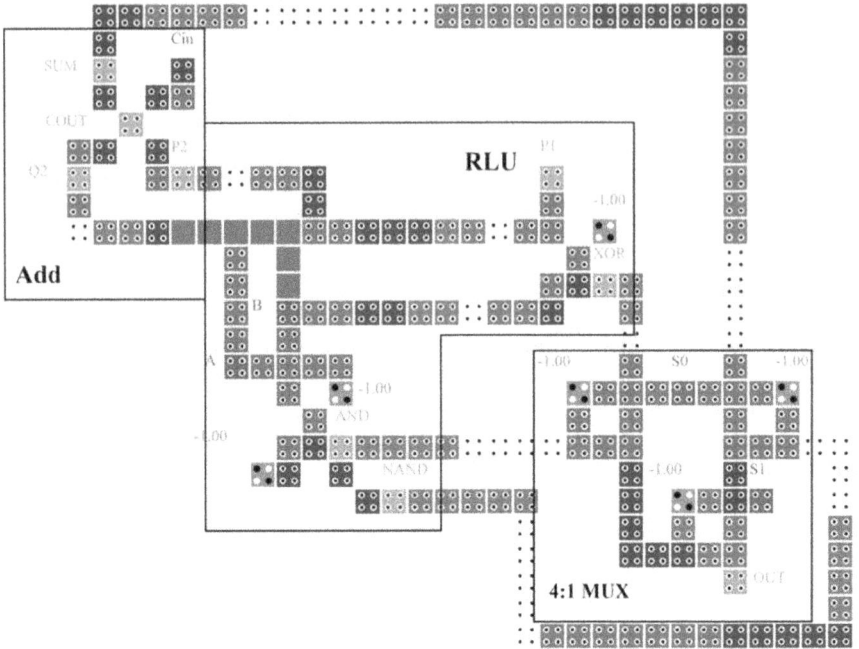

FIGURE 9.6   The presented first RALU.

TABLE 9.1
The operation table for 4:1 MUX in the first proposed RALU

S1	S0	Output Operation
0	0	AND
0	1	XOR
1	0	NAND
1	1	ADD

output of AND, NAND, XOR, and full adder is transferred to the input lines of the multiplexer. The output, OUT, is determined according to the selection lines, S0 and S1. This developed RALU composed of 209 cells, and 0.29 $\mu m^{2\ area}$.

9.3.3   THE SECOND PROPOSED RALU

The presented second RALU circuit is indicated in Figure 9.7. It includes the proposed RLU unit, full adder and multiplexer gates presented in Figures 9.5, in [30], and in [10], respectively.

The proposed RALU circuit is implemented in 3 layers. Table 9.2 displays the 4:1 MUX operation table.

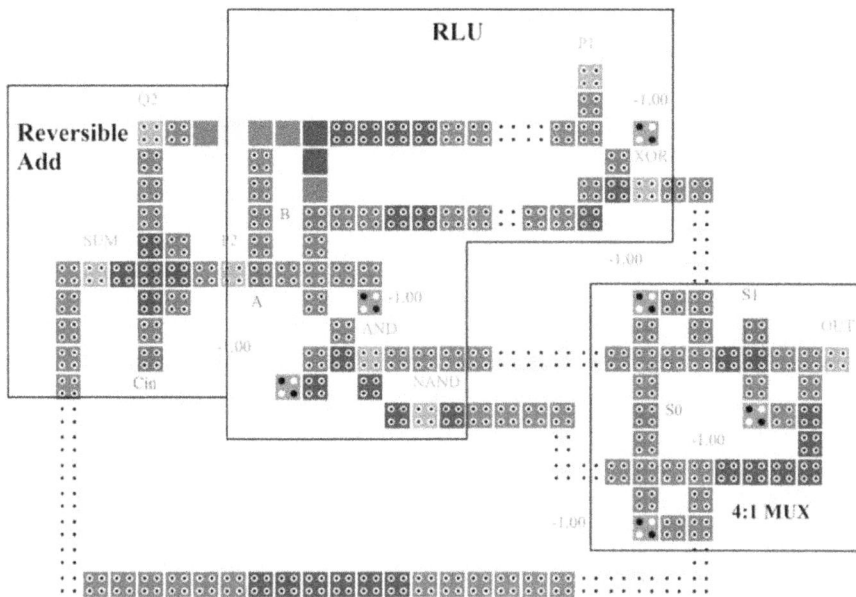

FIGURE 9.7   The presented second RALU.

TABLE 9.2
The operation table for 4:1 MUX in the second proposed RALU

S1	S0	Output Operation
0	0	ADD
0	1	NAND
1	0	XOR
1	1	AND

In this RALU, the inputs C, B, and A are designed in 1st layer. In addition, the 4:1 multiplexer is designed in such a way that the output, OUT, is shown in 1st layer. The 2nd layer is implemented as an interface to transfer information between the 1st layer and the 3rd layer. The full adder and the MUX are implemented in three layers. The output of AND, NAND, XOR, and full adder is transferred to the input lines of the multiplexer. The output, OUT, is determined according to the selection lines, S0 and S1. This developed RALU composed of 174 cells, and $0.24 \ \mu m^{2 \ area}$.

9.3.4 THE THIRD PROPOSED RALU

The presented third RALU is indicated in Figure 9.8. It includes the proposed RLU unit, full adder and multiplexer gates presented in Figures 9.5, in [30], and in [11], respectively.

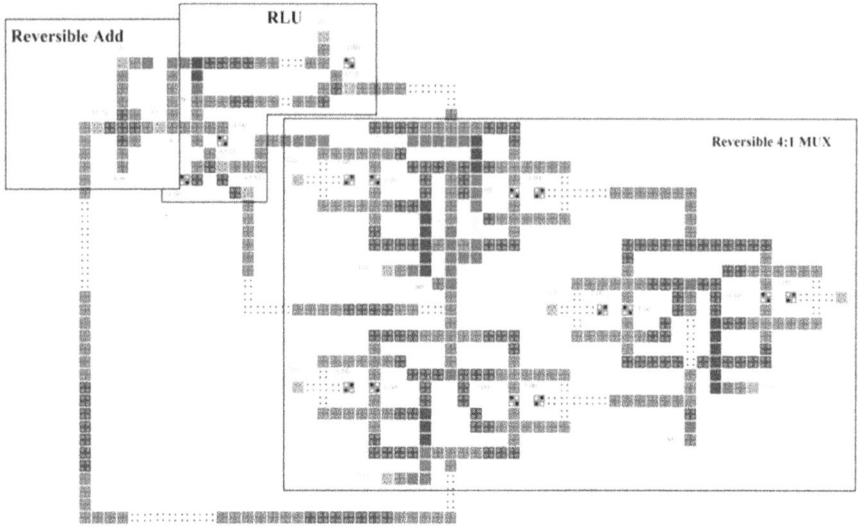

FIGURE 9.8   The presented third RALU.

TABLE 9.3
The operation table for 4:1 MUX in the third proposed RALU

S1	S0	Output Operation
0	0	XOR
0	1	AND
1	0	NAND
1	1	ADD

This circuit is implemented in three layers. Table 9.3 displays the 4:1 MUX operation table.

In this RALU, the inputs C, B, and A are designed in 1st layer. In addition, the 4:1 multiplexer is designed in such a way that the output, OUT, is shown in 1st layer. The 2nd layer is implemented as an interface to transfer information between the 1st layer and the 3rd layer. The full adder and the MUX are implemented in three layers. The output of AND, NAND, XOR, and full adder is transferred to the input lines of the multiplexer. The output, OUT, is determined according to the selection lines, S0 and S1. The developed RALU composed of 596 cells, and 0.96 $\mu m^{2\,area}$.

9.4   RESULTS AND COMPARISON

Simulation of RALU circuits are performed by QCADesigner tool version 2.0.3. In this chapter, the engine Bistable approximation is utilized for simulation. Table 9.4 indicates the simulation settings.

TABLE 9.4

The utilized parameters in the QCADesigner tool

Parameter	Value
Number of samples (RLU Unit)	12800
Number of samples (RALUs)	128000
Clock amplitude factor	2
Iterations maximum number in each sample	100
Radius of effect	65

9.4.1   SIMULATION RESULTS

9.4.1.1   Results of the Developed RLU

The results of the presented RLU are indicated in Figure 9.9.

In this figure B and A are inputs and AND, XOR, NAND, and P2 are outputs. According to the results, the outputs are correctly obtained. The latency of the presented RLU is 0.5 clock cycles.

9.4.1.2   Simulation Results of First Developed RALU

The results of the first developed RALU are indicated in Figure 9.10.

In this figure, B, A, Cin, S0, and S1 indicate the inputs and OUT and Cout indicate the outputs. According to the results of simulation, the outputs are correctly obtained. The latency of the presented RALU is 2.5 clock cycles.

9.4.1.3   Simulation Results of Second Developed RALU

The results of the second developed RALU are indicated in Figure 9.11.

In this figure, B, A, Cin, S0 and S1 indicate the inputs, and OUT indicates the output. According to the results of simulation, the output is correctly obtained. The latency of the presented RALU is 2.75 clock cycles.

9.4.1.4   Simulation Results of Developed Third RALU

The results of the presented third RALU are indicated in Figure 9.12.

In this figure, B, A, Cin, S0 and S1 indicate the inputs, and OUT indicates the output. According to the results of simulation, the output is correctly obtained. The latency of the developed RALU is five clock cycles.

9.4.2   COMPARISON

Table 9.5 shows the evaluation results including the number of operations, cell count, the number of required layers, area, latency, and cost.

Based on the results, which are presented in Table 9.5, the complexity or cell count of our first and second QCA RALU circuits is superior than other QCA RALU circuits. The area of the designed first and second QCA RALU circuits is superior than QCA RALU circuits in [16, 17, 18, 21, 22]. The only QCA RALU circuits that

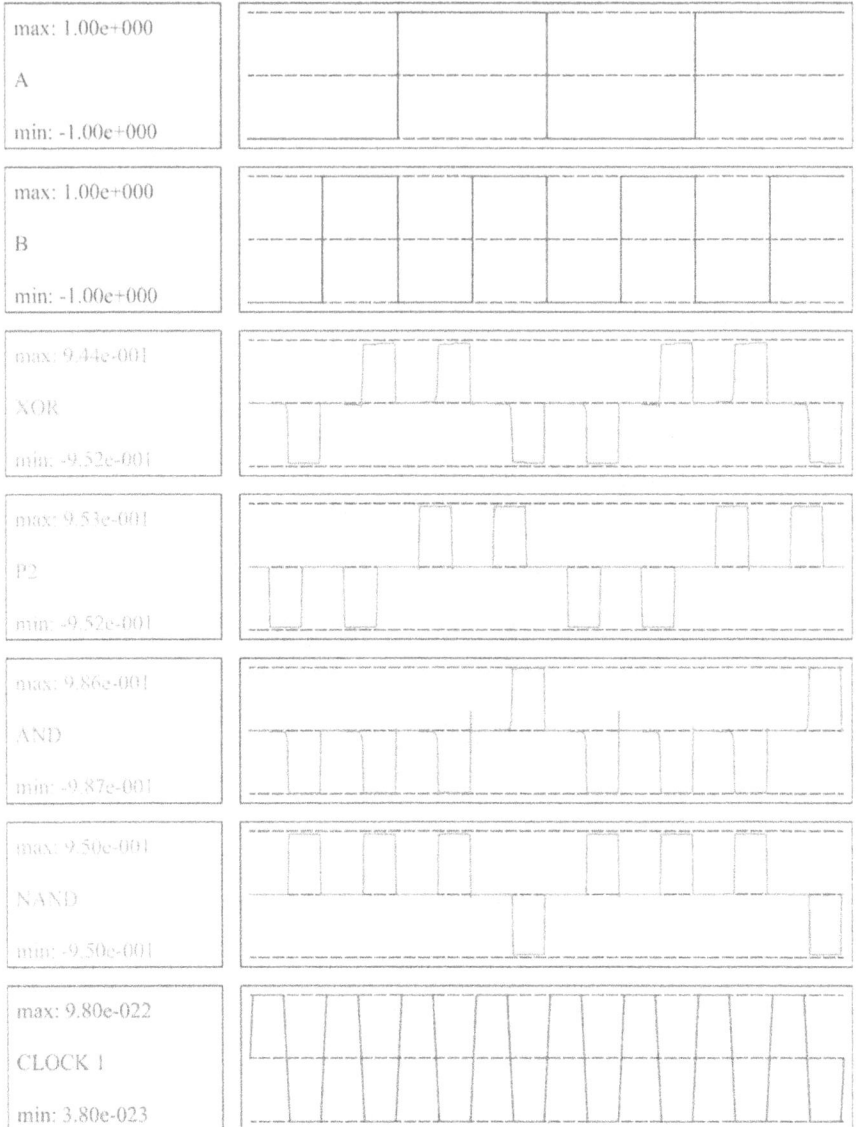

FIGURE 9.9   The results of the proposed RLU.

have slightly lower area compared to our developed RALU circuits are QCA RALU circuits presented in [19, 20]. It should be noted that these QCA RALU circuits are implemented in six and seven layers instead of three layers. The area of the designed first and second QCA RALU circuits is superior than QCA RALU circuits in [16, 17, 19, 20, 21, 22].

In addition, the complexity of the implemented third QCA RALU circuit is more suitable than QCA RALU designs in [12, 13, 15].

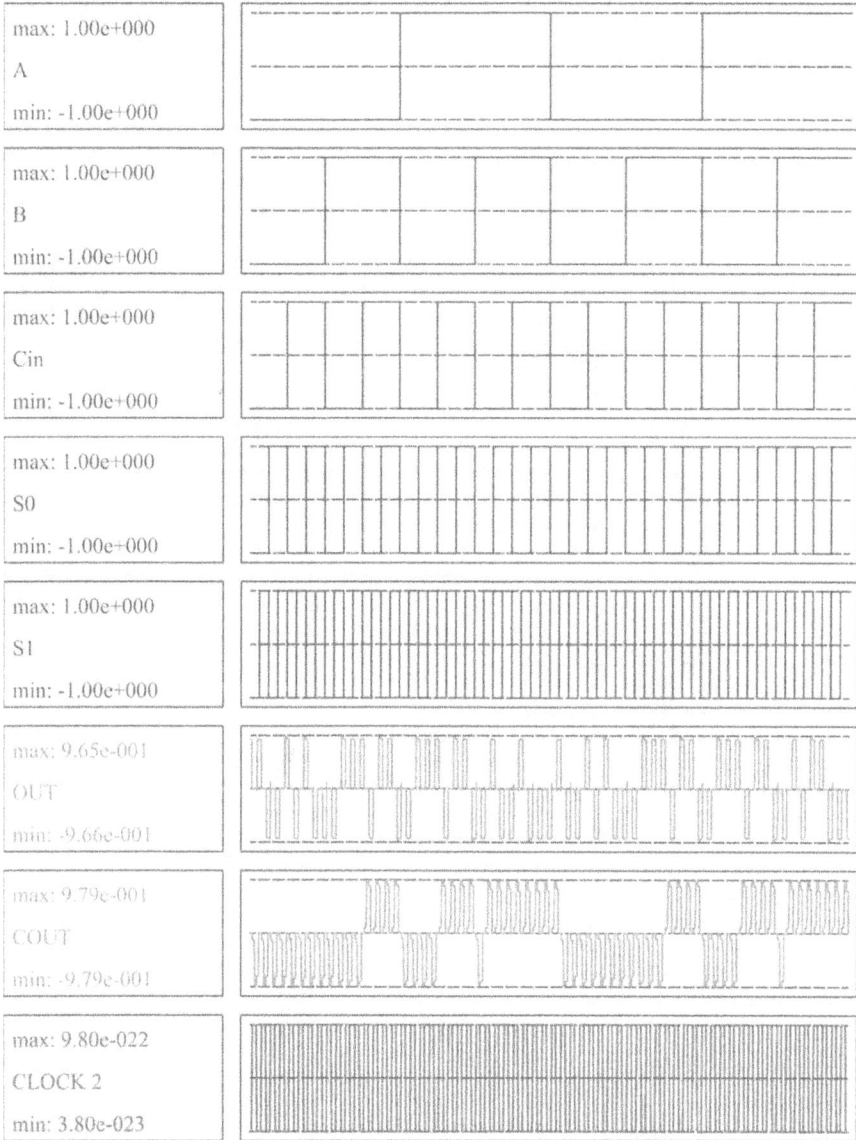

FIGURE 9.10 The results of simulation of the presented first RALU.

In addition, the cost of the QCA layout in Table 9.5 is computed using Equation (2) [2]:

$$Cost = Latency \times Area \tag{2}$$

Where Latency is in terms of clock cycles and Area is in in terms of μm^2. According to Table 9.5, the cost of the QCA layout of the first and second proposed RALU circuits are considerably improved compared to RALU circuits in [16, 17].

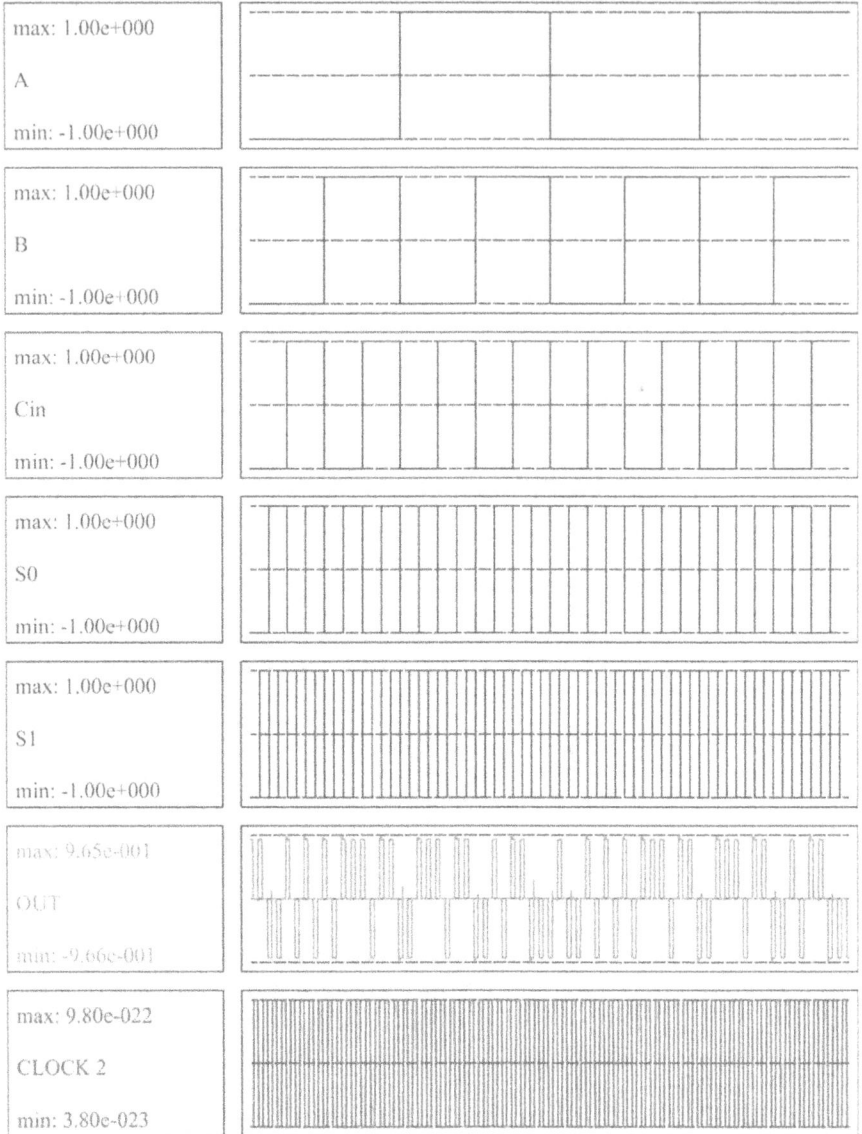

FIGURE 9.11   The results of simulation of the presented second RALU.

9.5   CONCLUSION

The QCA technology is a novel nanotechnology for implementing digital circuits that can be used for replacement of the CMOS technology. In this chapter, three RALU circuits were designed in three layers by using a new RLU, which was developed in this chapter. The proposed circuits contain four arithmetic and logical operations including addition, AND, NAND, and XOR. In the first proposed RALU, the aim

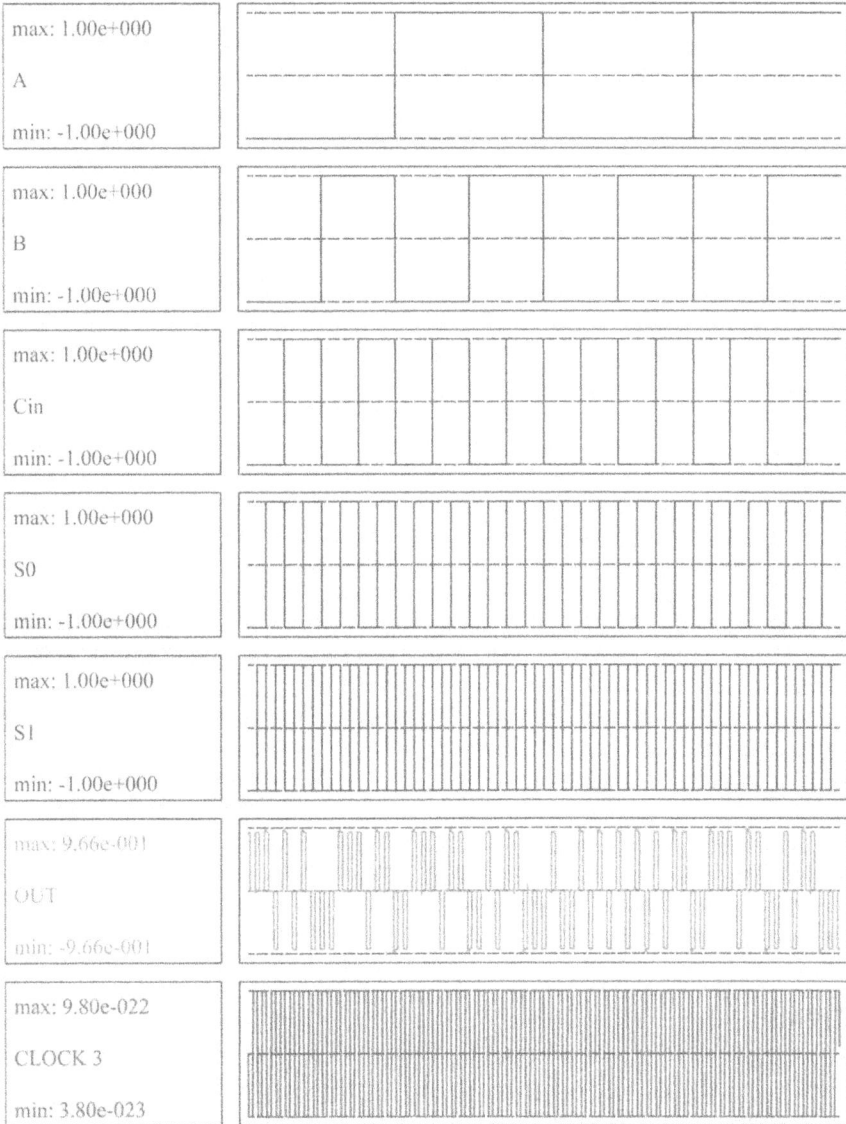

FIGURE 9.12 The results of simulation of the presented third RALU.

was to implement the simplest ALU using the basic gates. Then, we developed and optimized the first proposed circuit in the second and third proposed circuits in such a way that the proposed circuits have the least occupied area and cell count by using efficient path planning and component design. The main feature of the third circuit is in comparison with the first and second proposed circuits in the multiplexer unit, which is reversible in the third circuit and compared to other previous circuits, the

TABLE 9.5
Comparative table for the RALU circuits

Reference	Operations	Cell count	Area (μm^2)	Latency	layer	cost
[12]	8	1097	-	3.75	3	
[13]	16	2370	-	6	1	
[15]	16	670	-	4	3	
[16]	4	332	0.38	3	3	1.14
[17]	20	480	0.75	3.75	1	2.81
[18]	20	350	0.51	-	1	
[19]	4	230	0.1	3	6	
[20]	4	228	0.14	3.75	7	
[21]	20	625	1.34	3.75	1	
[22]	4	464	0.78	3	3	
Proposed 1	4	209	0.29	2.5	3	0.72
Proposed 2	4	174	0.24	2.75	3	0.66
Proposed 3	4	596	0.96	5	3	4.80

third RALU has better performance and efficiency. Our circuits are simulated using QCADesigner tool version 2.0.3. The first, second, and third developed QCA RALU circuits have 209, 174, and 596 cells. The developed first and second QCA RALU circuits show improvements in the design parameters of QCA circuit including the complexity, area, and latency compared to previous 1-bit RALU circuits. The third designed QCA RALU circuit has improvements compared with previous 1-bit RALU circuits in [13] regarding complexity. The cell count in the first and second developed RALU circuit is reduced compared with the developed RALU circuit in [16] by about 23% and 37%, respectively. The limitation of our circuits compared to other circuits is the number of arithmetic and logical operations, which includes only three logical operations and one arithmetic operation.

REFERENCES

[1] Rashidi, H. and Rezai, A., *Design of novel efficient multiplexer architecture for quantum-dot cellular automata*, Journal of Nano- and Electronic Physics, 2017, **9** (1):01012, 1–7.

[2] Roshany, H. and Rezai, A., *Novel efficient circuit design for multilayer QCA RCA*, International Journal of Theoretical Physics, 2019, **58**:1745–1757.

[3] Niknezhad Divshali, M. and Rezai, A. and Karimi, A., *Towards multilayer QCA SISO shift register based on efficient D-FF circuits*, International Journal of Theoretical Physics, 2018, **57**:3326–3339.

[4] Niknezhad Divshali, M. and Rezai, A. and Karimi, A., *Novel multilayer SISO shift register architecture in QCA technology and its usage in communications*, International Journal of Communication Systems, 2022, **35**:e5121.

[5] Salimzadeh, F. and Rasouli Heikalabad, S., *Design of a novel reversible structure for full adder/subtractor in quantum-dot cellular automata*, Physica B: Condensed Matter, 2019, **556**:163–169.

[6] Safoev, N. and Jeon, J., *Full adder based on quantum-dot cellular automata*, in Proceedings of International Conference of Trends in Engineering and Technology, 2017, 83–86.

[7] Balali, M. and Rezai, A., *Design of low-complexity and high-speed coplanar four-bit ripple carry adder in QCA technology*, International Journal of Theoretical Physics, 2018, **57**:1948–1960.

[8] Adelnia, Y. and Rezai, A., *A novel adder circuit design in quantum-dot cellular automata technology*, International Journal of Theoretical Physics, 2019, **58**:184–200.

[9] Rashidi, H., Rezai, A. and Soltany, S., *High-performance multiplexer architecture for quantum-dot cellular automata*, Journal of Computational Electronics, 2016, **15**:968–981.

[10] Alkaldy, E., Majeed, A., Zainal, M. and Nor, D., *Optimum multiplexer design in quantum-dot cellular automata*, Indonesian Journal of Electrical Engineering and Computer Science, 2020, **17**(1):148–155. arXiv preprint arXiv:2002.00360, 2020.

[11] Das, J., Purkayastha, T., and De, D., *Reversible nanorouter using QCA for nanocommunication*, Nanomaterials and Energy, 2016, **5**:28–42.

[12] Chaves, J., Silva, D., Camargos, V. and Neto, O., *Towards reversible QCA computers: Reversible gates and ALU*. In Proceeding of 6th IEEE Latin American Symposium on Circuits and Systems (LASCAS), 2015, 1–4.

[13] Sen, B., Dutta, M., Goswami, M. and Sikdar, B., *Modular design of testable reversible ALU by QCA multiplexer with increase in programmability*. Microelectronics Journal, 2014, **45**:1522–1532.

[14] Sasamal, T., Mohan, A. and Singh, A., *Efficient design of reversible logic ALU using coplanar quantum-dot cellular automata*, Journal of Circuits, Systems and Computers, 2018, **27**:1850021.

[15] Naghibzadeh, A. and Houshmand, M., *Design and simulation of a reversible ALU by using QCA cells with the aim of improving evaluation parameters*, Journal of Computational Electronics, 2017, **16**:883–895.

[16] Mirzajani Oskouei, S. and Ghaffari, A., *Designing a new reversible ALU by QCA for reducing occupation area*, The Journal of Supercomputing, 2019, **75**:5118–5144.

[17] Norouzi, M., Rasouli Heikalabad, S. and Salimzadeh, F., *A reversible ALU using HNG and Fredkin gates in QCA nanotechnology*, International Journal of Circuit Theory and Applications, 2020, **48**:1291–1303.

[18] Safaiezadeh, B., Mahdipour, E., Haghparast, M., Sayedsalehi, S., and Hosseinzadeh, M., *Novel design and simulation of reversible ALU in quantum dot cellular automata*, The Journal of Supercomputing, 2022, **78**:868–882.

[19] Roy, R., and Sarkar, S., *Physical design and verification of 3D reversible ALU by QCA technology*, Materials Today: Proceedings, 2022, In Press.

[20] Roy, R., and Sarkar, S., and Dhar, S., *Design and testing of a reversible ALU by quantum cells automata electro-spin technology*, The Journal of Supercomputing, 2021, **77**:13601–13628.

[21] Ahmadpour, S. Mosleh, M., and Rasouli Heikalabad, S., *The design and implementation of a robust single-layer QCA ALU using a novel fault-tolerant three-input majority gate*, The Journal of Supercomputing, 2020, **76**:10155–10185.

[22] Rahimpour Gadim, M., and Jafari Navimipour N., *A new three-level fault tolerance arithmetic and logic unit based on quantum dot cellular automata*, Microsystem Technologies, 2018, **24**:1295–1305.

[23] Bennett, C., *Logical reversibility of computation*, IBM Journal of Research and Development, 1973, **17**:525–532.

[24] Taheri, Z., Rezai, A. and Rashidi, H., *Novel single layer fault tolerance RCA construction for QCA technology*, Facta universitatis-series: Electronics and Energetics, 2019, **32**:601–613.

[25] Balali, M., Rezai, A., Balali, H., Rabiei, F. and Emadi, S., *Towards coplanar quantum-dot cellular automata adders based on efficient three-input XOR gate*, Results in physics, 2017, **7**:1389–1395.

[26] Sen, B., Nag, A., De, A. and Sikdar, B., *Multilayer design of QCA multiplexer*, In proceeding of 2013 Annual IEEE India Conference (INDICON), 2013, 1–6.

[27] Kim, S. and Swartzlander, E., *Parallel multipliers for quantum-dot cellular automata*, in proceeding of 2009 IEEE Nanotechnology Materials and Devices Conference, 2009, 68–72.

[28] Abutaleb, M., *Robust and efficient QCA cell-based nanostructures of elementary reversible logic gates,* The Journal of Supercomputing, 2018, **74**:6258–6274.

[29] Roohi, A., DeMara, R. and Khoshavi, N., *Design and evaluation of an ultra-area-efficient fault-tolerant QCA full adder*, Microelectronics Journal, 2015, **46**:531–542.

[30] Ahmad, P., Quadri, S., Ahmad, F., Bahar, A., Wani, G., and Tantary, S., *A novel reversible logic gate and its systematic approach to implement cost-efficient arithmetic logic circuits using QCA*, Data in brief 2017, **15**:701–708.

10 Stick Diagram Representation for MQCA-based Multiplexer

Vineet Jaiswal and Trailokya Nath Sasamal

10.1 INTRODUCTION

Magnetic Quantum-Dot Cellular Automata (MQCA) is a recent technology developed by Cowburn, Adeyeye, and Bland (1997). According to Semiconductor Industry Association (2016) leakage power is the main concern in any CMOS logic circuit as the number of transistors increases. Csaba, Lugli, Csurgay, and Porod (2005) illustrates that static power consumption is negligible in MQCA as all the logical information processes, propagates, and stores in bistable single-domain nanomagnets. MQCA devices work memory and logical devices, reducing the space required and increasing the data transfer speed. MQCA is the best choice for computing paradigms in harsh environments like radiation and heat, as stability is very high compared to semiconductor QCA devices (Siddiq et al. (2013)). Devices can be manufactured using standard litho-graphic techniques. With 10^{10} magnets switching 10^8 times/sec, it is estimated that they consume only about 0.1 W of power (Csaba et al. (2005)). These advantages make MQCA a promising contender for current logical devices alternative to conventional QCA and CMOS devices.

Nanomagnetic devices work on the ferromagnetic and antiferromagnetic coupling between adjacent nanodots and generate desired domain orientation. Antiferromagnetic coupling (AFC) occurs between nanodots when the easy axis are aligned adjacent to each other and Ferromagnetic coupling (FC) occurs when the hard axis are aligned adjacent to each other (Imre et al. (2006)). Shape anisotropy and external clock pulse decide the domain orientation, i.e. upward (logic value 1) or downward (logic value 0).

The hard axis is along the direction of the highest energy of polarization of the nanomagnets and perpendicular to the easy axis for implementing binary logic (Augustine, Behin-Aein, Fong, and Roy (2009)). It is possible to design logic gates driven by dipole coupling and use the direction of magnetization as the state variable instead of charge (Carlton, Emley, Tuchfeld, and Bokor (2008)). Therefore with the

DOI: 10.1201/9781003361633-10

help of domain orientation movement, combinational circuits are implemented, as demonstrated by Jaiswal and Sasamal (2022b); Sivasubramani, Mattela, Pal, Islam, and Acharyya (2018); Varga, Niemier, Csaba, Bernstein, and Porod (2013). Permalloy patterned nanodots exhibit minimum energy to switch from one orientation to another (Rodrigues et al. (2018)). In this chapter, we proposed a stick diagram representation of an NML-based multiplexer design that approximates the footprint for higher-bit applications. Furthermore, the stick diagram assists the research in higher-bit multiplexer design.

10.2   DESIGN OF NML-BASED MULTIPLEXER CIRCUIT

Multiplexer has applications in communication systems, arithmetic logic units (ALU), computer memory, and telephone network. To date, semiconductor QCA-based multiplexers have been designed widely (Khan and Mandal (2019); Rashidi and Rezai (2021)). MQCA-based multiplexer was implemented by Jaiswal and Sasamal (2022b) for the very first time. They implemented 2:1 multiplexer with the help of the traditional horizontal, vertical layout approach. Furthermore, they also demonstrated an area efficient compact design approach. All the layout designs and simulations were performed using a GPU-based MuMax3 micromagnetic simulator, which was developed by Vansteenkiste et al. (2014) at Ghent University. MuMax3 is very fast and has 100× speed compared to CPU-based micromagnetic simulators. Both CPU and GPU-based simulators determine the spin direction and magnetization of the adjacent nanomagnets using the Landau-Lifshitz equation (LLG) explained in equation (1).

$$\delta M(r,\ t)/\delta t = -\gamma M(r,\ t) \times H_{eff}(r,\ t) -$$
$$\delta\gamma/M_s\ [M(r,\ t) \times (M(r,\ t) \times H_{eff}\ (r,\ t))] \tag{1}$$

where $M(r,\ t)$ is the position and time dependent magnetization distribution, γ is the gyromagnetic ratio, α is damping constant respectively and H_{eff} is the pointwise effective field defined as:

$$H_{eff} = 1/\mu[\partial E(M)/\partial M]$$

where $E(M)$ is the Landau-Lifshitz energy and μ is the magnetic permeability. $E(M)$ is the summation of exchange energy, anisotropy energy, applied energy, and stray energy. We choose α value to be 0.5, and the default value of the gyromagnetic ratio is 1.7595×10^{11} rad/Ts in MuMax3. All the nanodots are made of permalloy, a soft ferromagnetic material with low coercive field nature and high permeability. In the past, most researchers used permalloy, a composition of 21.5% iron and 78.5% nickel for nanodots. The design and simulation parameters for MuMax3 are illustrated in Table 10.1, which is already used by Jaiswal and Sasamal (2022a); Sivasubramani et al. (2018). The solver uses the exchange stiffness and crystal anisotropy constants to calculate the exchange and anisotropy energies and provides the Landau-Lifshitz energy's E(M) component. The solver developed a steady-state spin moment in nanomagnets once the minimal energy-induced individual point-wise magnetization was reached in the time domain. MuMax3 does not require a layout editor for architectural design like OOMMF.

To date, no MQCA-based multiplexer has been designed by any researcher. So we implement a 2:1 multiplexer using these MQCA design techniques: vertical layout, horizontal layout, and compact layout (non-majority gate architecture). Horizontal and vertical layout designs work on the principle of three majority gates, as explained in equations (3) and (4). In contrast, compact layout design works on the principle of direct interaction, dominant nature, and coupling concept of slant edge nanomagnets.

$$Y = MG\ (MG(A,S,0),\ MG(B,S',0),\ 1) \tag{3}$$

$$Y = S'A + SB \tag{4}$$

Slant edge nanodots can generate independent input signals, requiring only one nanodot compared to the driver magnet configuration as explained by Varga (2013). The magnetization direction of slant-edge nanodots in the ground state depends on the nanodots' minimum energy and tilt planes. Exploiting this property of slant-edge nanomagnets reduces the number of nanodots and integrates this technology for the first time, taking our design into a new efficient domain. Slant-edge nanomagnets have energy landscapes according to their tilted side as shown in Figure 10.1. The figure shows energy (E) versus internal magnetization (H) for slant-edge nanomagnets with different slant-edge orientations. The thin line (H-axis) along the diagonal of the nanomagnet corresponds to the effective, easy axis. When the perpendicularly applied magnetic field does not align towards the energy maximum after the external magnetic field is removed, the nanomagnetic state rotates towards the closer of the

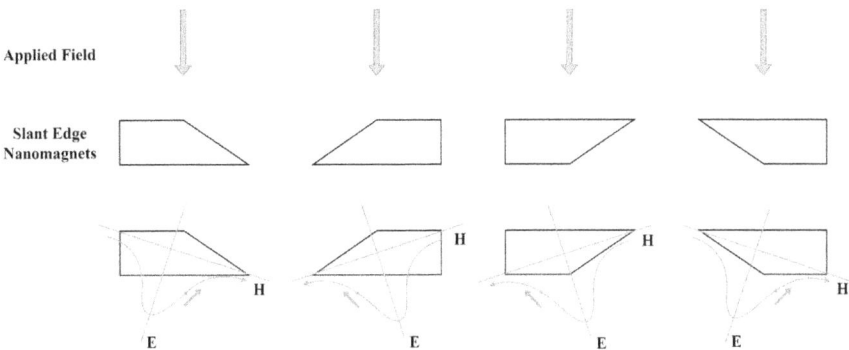

FIGURE 10.1 Energy landscape of slant edge nanomagnets after application of external magnetic field and removal of the same.

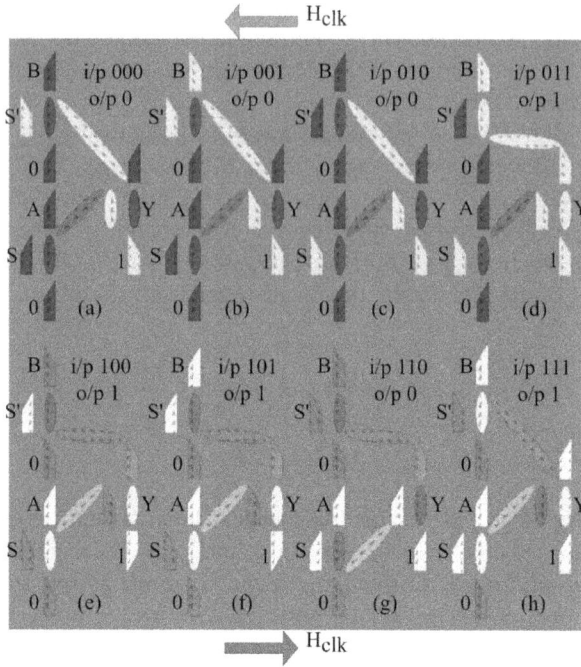

FIGURE 10.2 (a-h) MuMax3 simulation result of the nanomagnetic vertical layout (Jaiswal and Sasamal (2022b)).

two degenerate minimum energy (ground) states which is illustrated by Varga et al. (2013).

For the slant edge input nanomagnets, we used dimensions of $10 \times 30 \times 10$ nm^3. We used 10 nm thickness, 10 nm $\times$ 30 nm area, and 10 nm to 15 nm nanodot separation for antiferromagnetic and ferromagnetic coupling for the elliptical shape output nanomagnet. We designed a 2:1 multiplexer in MuMax3 micromagnetic simulator and performed a transient analysis until the domains of each nanomagnet were aligned to an overall minimum energy state. For 2:1 multiplexer, total energy, demagnetization energy, and exchange energy in the minimum energy state are 3.14×10^{-21} J, 4.14×10^{-21} J, and 8.16×10^{-24} J, respectively. MuMax3 simulation results for vertical and horizontal designs are illustrated in Figures 10.2 and 10.3 respectively. Figure 10.4 corresponds to the proposed compact layout design and their simulation output using the MuMax3 micromagnetic simulator.

10.3 STICK DIAGRAM MODEL FOR HIGHER BIT MULTIPLEXER

2:1 multiplexer was realized using vertical, horizontal, and compact layout designs. The vertical and horizontal layout uses three majority gates, whereas the compact layout uses only one non-majority gate structure. To show the furthermore application of 2:1 multiplexer we have implemented an 8:1 multiplexer and simulation output

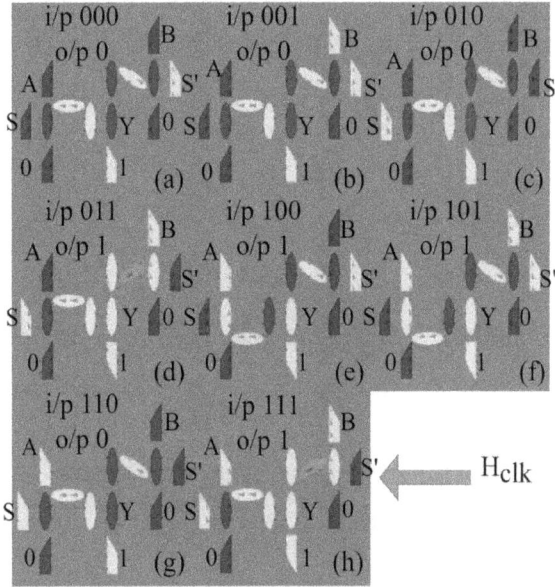

FIGURE 10.3 (a-h) MuMax3 simulation result of the nanomagnetic horizontal layout (Jaiswal and Sasamal (2022b)).

for 00000000 and 11111111 input combination shown in Figure 10.5 (a, b). For 8:1 multiplexer, we use different sizes of inclined interconnect (IC) nanomagnets as in vertical and horizontal designs. As the size of interconnecting nanomagnets changes, there is difficulty in designing a higher-bit multiplexer. Therefore, to overcome this problem, we use an array of antiferromagnetically coupled nanomagnets for fan-out and interconnection between 2:1 multiplexer designs. The limitation of using an array of nanomagnets is that the number of nanomagnets increases compared to the inclined interconnect structure. 2:1 multiplexer layout design and simulation output is shown in Figure 10.4, which is derived from one non-majority gate structure. Here A and B are inputs, S is the select line, and Y is the output. In addition, an 8:1 multiplexer was implemented, as shown in Figure 10.5. Here, we use seven non-majority blocks of 2:1 multiplexer. The number of non-majority blocks can be derived as $(2^n - 1)$ for higher order 2^n: 1 or N: 1 multiplexer where $N = 2^n$.

The number of non-majority blocks are summarized in Table 10.2 which is helpful in the higher-bit application. Here we proposed a model which will approximate the footprint for higher-bit applications. Stick diagram representation gives the impression of the physical implementation of the faster and higher bit multiplexer and is useful in the routing and placement stage. The footprint of the higher-order multiplexer recapitulated in Table 10.2 is provided in terms of the width (W_{cd}) and length (L_{cd}) of the non-majority logic blocks, where W_d illustrates the width of nanodots and the distance between horizontally placed nanodots and L_d refers to the length of nanodots and the distance between ferromagnetically coupled nanomagnets. In our proposed compact layout design of 2:1 multiplexer

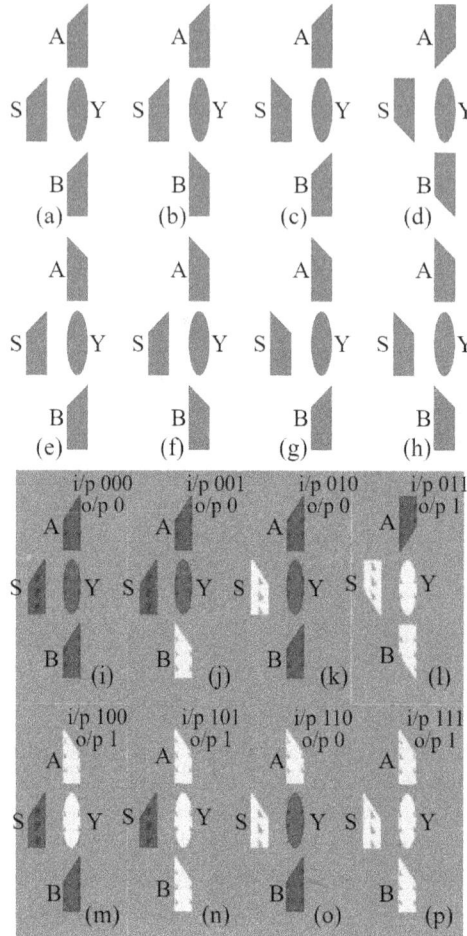

FIGURE 10.4 (a-h) Proposed compact layout design, (i-p) MuMax3 simulation output of the proposed design (Jaiswal and Sasamal (2022b)).

L_{cd} is 100 nm and W_{cd} is 30 nm. Therefore total area of the 2:1 multiplexer compact design is 0.003 μm^2. The overall area of higher order multiplexer is $2^{n-1}L_{cd} + (2^{n-1} - 1)L_d \times nW_{cd} + (n - 1)W_d$.

For 4:1 multiplexer W_d 50 nm and L_d is 40 nm. The footprint for 4:1 multiplexer is shown in Figure 10.6 and area occupied is $(2L_{cd} + L_d \times 2W_{cd} + W_d)$. The layout design of 8:1 multiplexer is shown in Figure 10.7 and area occupied is $(4L_{cd} + 3L_d \times 3W_{cd} + 2W_d)$. For 16:1 multiplexer area occupied is $(8L_{cd} + 7L_d \times 4W_{cd} + 3W_d)$. Thus generalized formula to calculate the area of 2^n: 1 multiplexer is $2^{n-1}L_{cd} + (2^{n-1} - 1)L_d \times nW_{cd} + (n - 1)W_d$.

The stick diagram proposed in nanomagnetic logic simplifies point-level schemes and furnishes supplemental mask layer details for structuring the nanomagnets of the higher-order systems (Oraon and Rao (2019)). For nanomagnetic logic,

TABLE 10.1
MuMax3 design and simulation parameters

Design and simulation parameters	
Material	Permalloy
Ms (saturation magnetization)	800×10^3 A/m
A (exchange stiffness constant)	13×10^{-12} J/m
α (damping coefficient)	0.5
Temperature	300 K
Grid size	(128, 256, 1)
Cell size	$0.5 \times 0.5 \times 0.5$ nm^3
Slant edge nanomagnet size	$10 \times (15, 30) \times 10$ nm^3
Ellipsoidal nanomagnet size	$10 \times 30 \times 10$ nm^3
Horizontal distance	10 nm
Vertical distance	5 nm

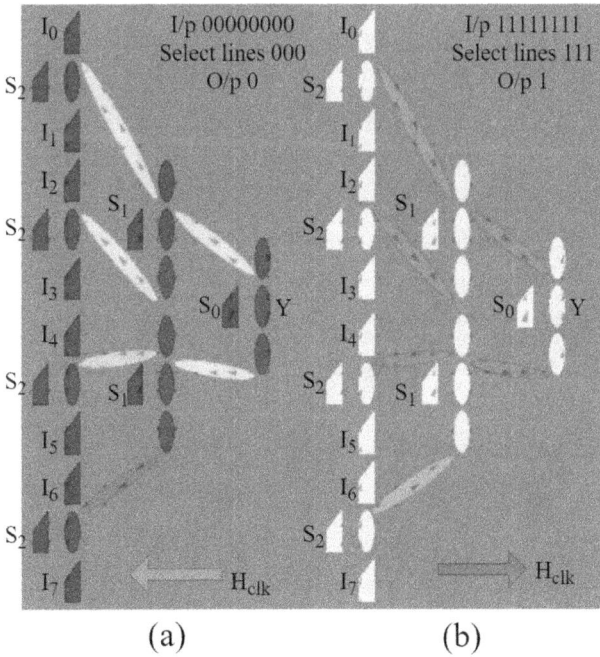

(a)   (b)

FIGURE 10.5 (a, b) Simulation output of the 8:1 multiplexer for 00000000 & 11111111 input combination (Jaiswal and Sasamal (2022b)).

the stick diagram is demonstrated by orthogonal lines, with the horizontal line showing the nature of the antiferromagnetically coupled nanomagnets and the vertical line showing ferromagnetic behavior between the adjacent nanomagnets. Electrically isolated nanomagnets are magnetically coupled, so lines on a stick

TABLE 10.2
Number of non-majority blocks to represent N bit multiplexer

N bit Multiplexer ($N = 2^n$)	Number of non-majority blocks ($2^n - 1$)	Length	Width
2:1	1	L_{cd}	W_{cd}
4:1	3	$2L_{cd} + L_d$	$2W_{cd} + W_d$
8:1	7	$4L_{cd} + 3L_d$	$3W_{cd} + 2W_d$
16:1	15	$8L_{cd} + 7L_d$	$4W_{cd} + 3W_d$
2^n: 1	$2^n - 1$	$2^{n-1}L_{cd} + (2^{n-1} - 1)L_d$	$nW_{cd} + (n - 1)W_d$

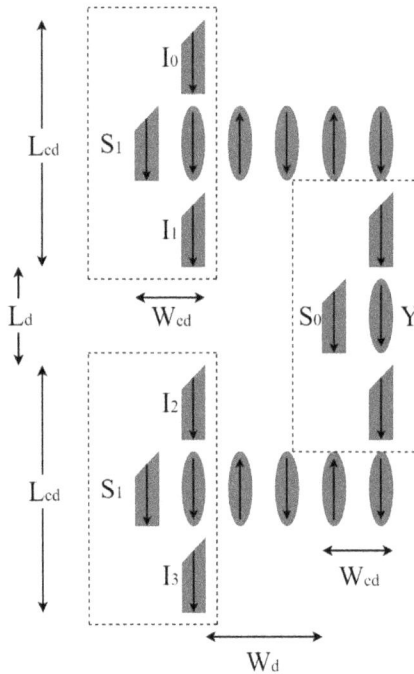

FIGURE 10.6   Layout design of magnetic logic based 4:1 multiplexer using 2:1 multiplexer.

diagram show coupling between horizontally or vertically arranged nanodots. The red intersection lines indicate that the organized antiferromagnetic or ferromagnetic coupling is specified to one plane level. A stick diagram representation of the 4:1 multiplexer circuit which is illustrated in Figure 10.8 containing mainly non-majority blocks is demonstrated by vertical and horizontal lines showing the placement of nanomagnets along the easy and hard axes. The center of the

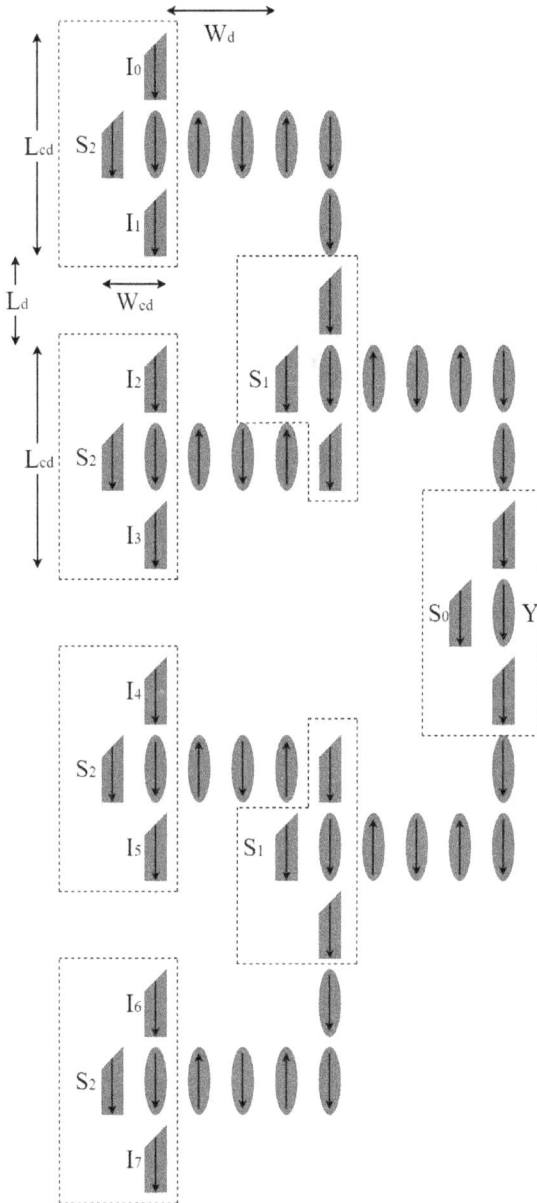

FIGURE 10.7 Layout design of magnetic logic based 8:1 multiplexer using 2:1 multiplexer.

intersection point of the orthogonal lines represents the output logic of the non-majority block.

Figure 10.9 demonstrates a stick diagram representation of the 8:1 multiplexer. In stick diagram representation the non-majority compact layout block is replaced by

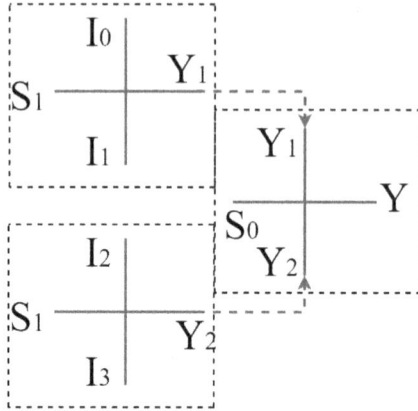

FIGURE 10.8   Stick diagram representation of 4:1 multiplexer.

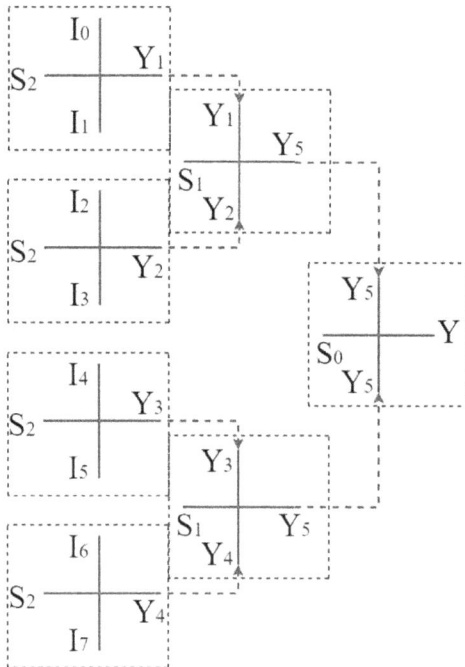

FIGURE 10.9   Stick diagram representation of 8:1 multiplexer.

intersecting orthogonal lines. Figure 10.10 shows a peripheral overview, including signal generation and propagation sketches for the N:1 multiplexer subsystem, derived from stick diagram representations of the 4:1 and 8:1 multiplexer subsystems. A higher-order N:1 multiplexer is shown in a dotted blue line box.

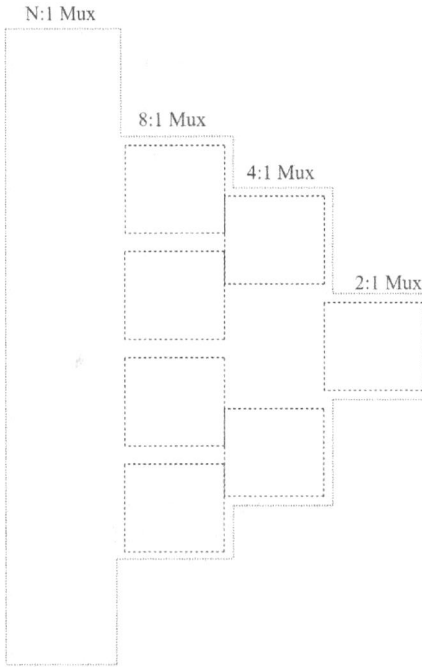

FIGURE 10.10   Stick diagram representation of N:1 multiplexer.

10.4   CONCLUSION

An NML-based 2:1 multiplexer layout was designed and simulated in the MuMax3 simulator. The compact layout design of the 2:1 multiplexer was extended to N-bit multiplexer with the help of the repeated use of the above. We can approximate the higher-order circuit's length, width, and area occupancy footprint. Furthermore, for a fast approximation of the higher-bit combination circuit, a stick diagram model was derived using the compact model design. The Stick diagram model simplifies the routing and planning phases of the higher-order multiplexer. A logic generation and propagation model specific to the design of N-bit high-order multiplexer subsystems were developed to share design topography and logic propagation information. A stick diagram representation is proposed using the horizontal and vertical lines, providing a fast footprint approximation of the MQCA subsystem. It is considered a technology for ultra-low power and non-real-time applications.

REFERENCES

Augustine, C., Behin-Aein, B., Fong, X., & Roy, K. (2009). A design methodology and device/circuit/architecture compatible simulation framework for low-power magnetic quantum cellular automata systems. *Proceedings of the Asia and South Pacific Design Automation Conference, ASP-DAC*, 847–852.

Carlton, D. B., Emley, N. C., Tuchfeld, E., & Bokor, J. (2008). Simulation studies of nanomagnet-based logic architecture. *Nano Letters*, *8* (12), 4173–4178.

Cowburn, R. P., Adeyeye, A. O., & Bland, J. A. (1997). Magnetic switching and uniaxial anisotropy in lithographically defined anti-dot Permalloy arrays. *Journal of Magnetism and Magnetic Materials*, *173* (1–2), 193–201.

Csaba, G., Lugli, P., Csurgay, A., & Porod, W. (2005). Simulation of power gain and dissipation in field-coupled nanomagnets. *Journal of Computational Electronics*, *4* (1–2), 105–110.

Imre, A., Csaba, G., Ji, L., Orlov, A., Bernstein, G. H., & Porod, W. (2006). Majority logic gate for magnetic quantum-dot cellular automata. *Science*, *311* (5758), 205–208.

Jaiswal, V., & Sasamal, T. N. (2022a). Novel approach for the design of efficient full adder in MQCA. *The Journal of Supercomputing* (0123456789), *79*(7), 7900–7915. Retrieved from https://doi.org/10.1007/s11227-022-04989-0

Jaiswal, V., & Sasamal, T. N. (2022b). A Novel Approach to Design Multiplexer using Magnetic Quantum-Dot Cellular Automata. *IEEE Embedded Systems Letters*, 1.

Khan, A., & Mandal, S. (2019). Robust multiplexer design and analysis using quantum dot cellular automata. *International Journal of Theoretical Physics*, *58* (3), 719–733.

Oraon, N., & Rao, M. (2019). Stick diagram representation for nanomagnetic logic based combinational circuits. *Proceedings of the IEEE Conference on Nanotechnology*, *2018-July*, 2–5.

Rashidi, H., & Rezai, A. (2021). Design of novel multiplexer circuits in QCA nanocomputing. *Facta universitatis–series: Electronics and Energetics*, *34* (1), 105–114.

Rodrigues, D. C., Klautau, A. B., Edström, A., Rusz, J., Nordström, L., Pereiro, M.,.... Eriksson, O. (2018). Magnetic anisotropy in permalloy: Hidden quantum mechanical features. *Physical Review B*, *97* (22), 1–6.

Semiconductor Industry Association. (2016). International technology roadmap for semiconductors 2.0, 2015 Edition Executive Report. *Itrpv*, *0* (March), 1–37.

Siddiq, M. A., Niemier, M. T., Csaba, G., Orlov, A. O., Hu, X. S., Porod, W., & Bernstein, G. H. (2013). A nanomagnet logic field-coupled electrical input. *IEEE Transactions on Nanotechnology*, *12* (5), 734–742.

Sivasubramani, S., Mattela, V., Pal, C., Islam, M. S., & Acharyya, A. (2018). Shape and positional anisotropy based area efficient magnetic quantum-dot cellular automata design methodology for full adder implementation. *IEEE Transactions on Nanotechnology*, *17* (6), 1303–1307.

Vansteenkiste, A., Leliaert, J., Dvornik, M., Helsen, M., Garcia-Sanchez, F., & Van Waeyenberge, B. (2014). The design and verification of MuMax3. *AIP Advances*, *4* (10), 0–22.

Varga, E. (2013). *Experimental study of novel nanomagnet logic devices* (Unpublished doctoral dissertation).

Varga, E., Niemier, M. T., Csaba, G., Bernstein, G. H., & Porod, W. (2013). Experimental realization of a nanomagnet full adder using slanted-edge magnets. *IEEE Transactions on Magnetics*, *49* (7), 4452–4455.

11 Fully Depleted Planar Bi-layer Junctionless Transistor for Future Technology Node

Gaurav Saini[1] and Trailokya Nath Sasamal[1]
[1]Department of ECE, NIT Kurukshetra, Haryana, India

11.1 INTRODUCTION

Today, the most common and challenging fabrication step of a semiconductor device is the formation of junctions. At technology nodes below 20 nm, the formation of super-steep doping profiles across source–channel–drain regions imposes severe challenges (Colinge et al. 2010). To maintain the scaling pace in the coming future, there is a need for junctionless devices like junctionless transistors (JLT) proposed by many researchers (Colinge et al. 2010; Lilienfeld (1925); Colinge et al. 2011). Various structures of JLTs have been proposed such as tri-gate nanowire JLT on silicon on insulator (SOI) (Colinge et al. 2010), junctionless multigate field-effect transistor on SOI (Lee et al. 2009), bulk planar JLT (BPJLT) (Gundapaneni, Ganguly, and Kottantharayil, 2011), planar JLT with non-uniform channel doping (Mondal, Ghosh, and Bal 2013), tri-gate JLTs with dual-k sidewall spacers (Saini G. and Choudhary S. 2016a; Saini G. and Choudhary S. 2016b). Among all JLT devices, single gate planar JLT in SOI and BPJLT are of interest because of their compatibility with standard planar CMOS process flow. The key factors that affect JLT performance are silicon film thickness, gate work function and silicon film doping concentration at a given gate length. For SOI JLT, silicon film thickness should be small enough so that it can be depleted by the workfunction difference of metal gate and doped silicon film under the gate (Colinge et al. 2010). In order to reduce SCEs of nanowire junctionless transistor, silicon film thickness and/or width should be at least twice as small as the channel length (Colinge et al. 2011; Choi, Moon, Kim, Duarte and Choi 2011). Another important issue is the requirement of high gate work function ~5.5 eV for extremely scaled SOI JLTs (Gundapaneni et al. 2011). In single-gate SOI-JLT, the main reason for leakage is that the gate control decreases with an increase in channel depth and carriers are less depleted at the bottom of the channel and more depleted near the gate (Mondal et al. 2013; Sallese, Chevillon, Lallement, Iniguez and Pregaldiny 2011). During the OFF state, leakage current flows through the bottom of the channel and degrades the ON-OFF current ratio. The aforementioned problem has been addressed by various researchers. They have proposed two different single-gate JLT structures, i.e. Bulk Planar Junctionless Transistor (BPJLT) (Gundapaneni et al. 2011) and Gaussian Doped SOI Junctionless Transistor (GDP-SOIJLT) (Mondal et al. 2013). However, the problem is that both devices require high gate workfunction values, above 5 eV, to maintain a good ON-OFF current ratio and device scalability. The requirement of high gate work function is a serious and technologically challenging limitation.

Therefore, it is interesting to search for a fully depleted JLT device structure that requires a small gate work function under off-state conditions for future technology nodes. To address this problem, we propose a single gate SOI-JLT with a bi-layer structure (BJLT). The advantages of the proposed BJLT structure over the other JLT structures are its ability to reduce leakage current even with midgap gate work function (4.6 eV) gate materials, good ON-OFF current ratio and better scalability. In Section 11.2, the proposed device structure is investigated, an analysis of the results is presented in Section 11.3 and the conclusion of the study is presented in Section 11.4.

11.2 DEVICE STRUCTURE AND OPERATION

Figure 11.1 shows the n-channel UJLT and BJLT structures realized using Sentaurus TCAD tool suite (Synopsys Inc. 2020). In Figure 11.1 (b), a p-type silicon layer of 10 nm is formed on top of a thick buried oxide (BOX) layer of 20 nm. The upper half portion of the p-layer is transformed into an n-layer by doping (Gaussian profile). The device layer (n-type) is uniformly doped along the channel length (in the horizontal plane) and non-uniformly doped in the vertical plane.

In this study, we investigate the n-channel single gate unilayer Gaussian Doping Profile (GDP) SOI-JLT, we call it a unilayer junctionless transistor (UJLT), and compare it with the proposed BJLT structure. The term UJLT and BJLT means Gaussian doping profile UJLT and BJLT unless stated otherwise. In order to perform this comparative study, the simulation parameters of both devices are chosen in confirmation with the parameters used by other reported works and are presented in Table 11.1. All parameters for both devices are kept the same except device layer(s) thickness. In the

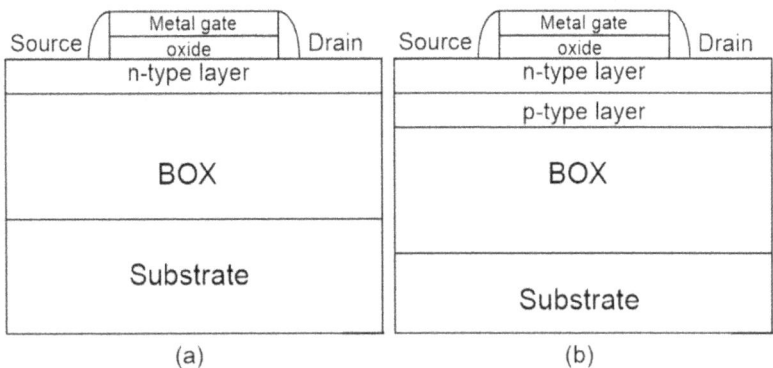

FIGURE 11.1 Schematic representation of the (a) unilayer n-channel junctionless transistor (UJLT) (b) proposed bilayer n-channel junctionless transistor (BJLT).

TABLE 11.1
Parameters used for the device simulation

Parameter	Value
Device layer thickness (T_{si})	5 nm n-type for UJLT, 10 nm (5 nm n-type and 5 nm p-type) for BJLT
Donor doping (N_d)	1.5×10^{19} /cm^3
Acceptor doping (N_a)	5×10^{18} /cm^3
EOT of gate dielectric (T_{ox})	1 nm
Channel length (L_g)	10 nm to 50 nm
Drain Voltage (V_{DD})	1.0 V
Gate Work function (Φ_m)	4.8 eV

case of UJLT and BJLT structures, the peak doping concentration is 1.5×10^{19}/cm^3 and the standard deviation is 3.5 nm.

11.2.1 EQUILIBRIUM AND NON-EQUILIBRIUM TRANSPORT

Without any external bias applied to BJLT, the n-layer under the gate gets depleted due to the metal gate-silicon work function difference and diffusion of majority carriers across p-n junction. However, the n-layer towards the source and drain side gets depleted due to the diffusion of the majority carriers across p-n junction. On the other hand, p-layer under the source and drain electrode is weakly inverted due to the formation of depletion region at the p and n-layer junction and also due to silicon-BOX interfacial boundary conditions (Kencke et al. 1999). A similar phenomenon was reported by Kencke et al. (1999) and is called Zero-bias Internal Barrier Lowering (ZIBL). In the proposed BJLT structure, the p-layer under the gate region is at first weakly inverted due to ZIBL and p-n junction depletion effect, and then depleted back due to the difference in the work function of gate material and doped silicon layer (see Figure 11.2). Under equilibrium, both device layers under the gate will be fully depleted and hence we call it a fully depleted BJLT. At zero gate bias, both n and p-type layers get fully depleted and hence smaller subthreshold leakage current flows from drain to the source terminal. As a positive voltage is applied to the gate terminal, the n-layer comes out of depletion and the p-layer reverts back to its zero bias equilibrium state, i.e. weak inversion state. Therefore, there is a flow of current from drain to source through both the n-layer and weakly inverted p-layer. To justify our intuition, we investigate this device structure and the results are shown in Figures 11.2–11.4.

From Figure 11.2, we notice that in the OFF state the BJLT device layers (both n-and p-type layers) are depleted very well with a 4.8 eV gate work function.

This is the main advantage of this device structure and hence it can relax the technologically challenging device design constraints like the requirement of high workfunction gate material and/or thin active layer on the top of BOX to ensure less subthreshold leakage with high ON-OFF current ratio. As seen in Figure 11.2, in the ON state, the conduction band energy of the n-layer is lower than that of p-layer. It means that the depleted p-layer is now weakly inverted and the depleted n-layer is now out of depletion.

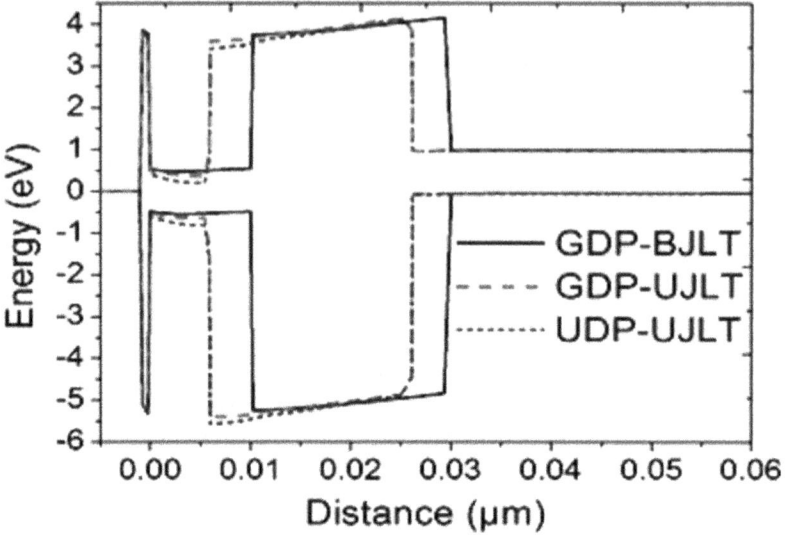

FIGURE 11.2 Energy band diagram of n-channel Gaussian doping profile (GDP) BJLT, GDP-UJLT and uniformly doping profile (UDP) UJLT at 20 nm gate length along a vertical cross-section through the middle of the channel in OFF state ($V_{GS} = 0$ V, $V_{DS} = 1$ V) with a gate work function of 4.8 eV.

11.3   RESULTS AND DISCUSSION

The simulations were performed using the density-gradient quantum correction model, Schokley-Read-Hall model, Auger recombination model and bandgap narrowing model. Figure 11.3 shows the scaling capabilities of a UJLT and BJLT for channel lengths from 50 nm down to 10 nm with the gate work function of 4.8 eV. As the channel length of UJLT is reduced, the ON-OFF current ratio reduces from 2.47×10^8 to 3.37×10^4, respectively. We can further improve the ON-OFF current ratio of UJLT by a super steep doping profile in the vertical direction as suggested in [8]. However, the requirement of a high value of gate work function and thin silicon layer remains unchanged for highly scaled devices. As the channel length of BJLT is reduced, the ON-OFF current ratio reduces from 6.07×10^9 to 1.92×10^5 respectively, as seen in Figure 11.3.

Thus, we find that the ON-OFF current ratio of a BJLT is ~1 order higher than the ON-OFF current ratio of a UJLT with gate work function of 4.8 eV at 20 nm gate length. It is interesting to note here the percentage reduction in ON and OFF state current from UJLT to BJLT is 52% and 93.09% respectively (at 20 nm gate length and gate work function of 4.8 eV). Hence, we find that BJLT can work efficiently with lower values of gate work function ranging from 4.6 eV to 4.8 eV. It is clearly visible from Figure 11.3 and Figure 11.4 (a) that BJLT has a superior ON-OFF current ratio and better scalability in comparison to conventional single gate JLTs. Validation of simulation results is carried out through the calibration of simulation setup with experimental data (Vinet et al. 2005) shown in Figure 11.4 (b).

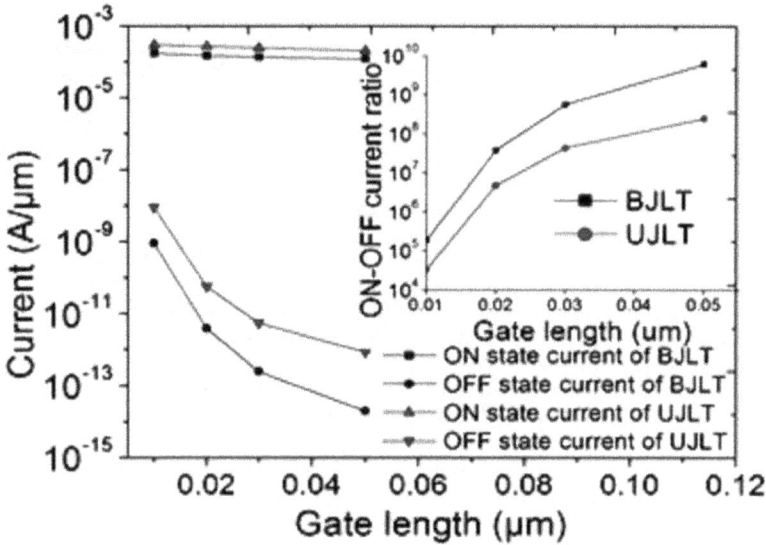

FIGURE 11.3 Variation of ON and OFF state current along the gate length for n-channel BJLT and UJLT with the gate work function of 4.8 eV.

Table 11.2 shows the comparison of analog performance between n-channel UJLT and BJLT at a drain current of $I_{ds}=10$ µA/µm. The values of g_m, g_{ds} and C_{gg} are found to be more in BJLT than UJLT. The percentage of the increment (14%) in g_m is more than the C_{gg} in BJLT. Therefore, an increment of 5.8% is recorded in the value of f_T $(=g_m/2\pi C_{gg})$ for BJLT. A reduction in the value of V_{EA} is observed in BJLT structure because of its low output resistance. Moreover, the ratio (C_{gs}/C_{gd}) is found less in BJLT than UJLT structure. It shows parasitic feedback capacitance (Dambrine et. al. 2003) will be slightly more in BJLT due to its bi-layered structure.

11.4 CONCLUSION

In this work, a planar single-gate SOI junctionless transistor is proposed which consists of two semiconductors layers instead of one as in conventional JLTs. A 2-D numerical simulation study was performed with non-uniform Gaussian doping in the device's active layers. Simulation studies suggest an improvement in the ON-OFF current ratio of the proposed structure with improved scalability. The major advantage of the proposed device is that one can use lower gate work function materials to operate the device in the desired mode of operation with acceptable values of performance parameters. Moreover, the BJLT structure can provide a 13% improvement in transconductance (g_m) and 5.8% unit gain cut-off frequency (f_T) at the cost of low output impedance. However, we also understand that the design of this device imposes fabrication challenges to build two device layers with desired doping profile and concentration.

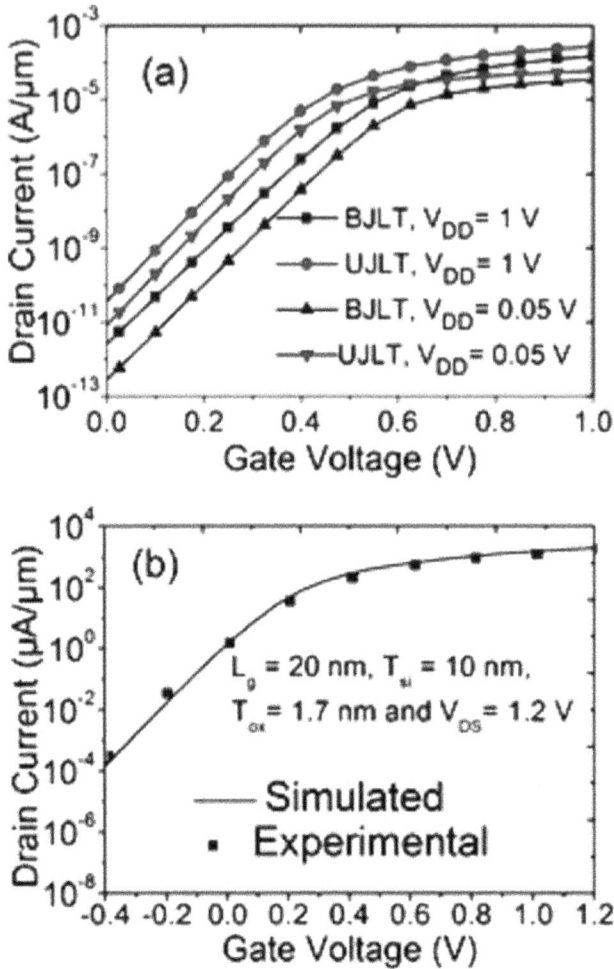

FIGURE 11.4 (a) I_{DS}-V_{GS} characteristic of n-channel BJLT and UJLT structures at 20 nm gate length and (b) comparison of simulated results and published experimental data.

TABLE 11.2
Comparison of analog/RF metrics of n-channel UJLT and BJLT at I_{ds} = 10 $\mu A/\mu m$ under matched off-state current of 1×10^{-11} $\mu A/\mu m$

Parameters	UJLT	BJLT
f_T (GHz)	138	146
A_{V_0} (dB)	64.5	63
V_{EA} (V)	0.78	0.57
C_{gg} (fF/μm)	0.24	0.26
g_m (μS/μm)	215	245
g_{ds} (μS/μm)	0.13	0.17
C_{gs}/C_{gd}	3.40	3.14

REFERENCES

Choi, S. J., Moon, D., Kim, S., Duarte, J. P., and Choi, Y. K. (2011), 'Sensitivity of Threshold Voltage to Nanowire Width Variation in Junctionless Transistors', *Electron Device Letters, IEEE*, 32(2), 125–127.

Colinge, J. P., Ferain, I., Kranti, A., Lee, C. W., Nima, D. A., Razavi, P., Yan, R. and Yu, R. (2011), 'Junctionless Nanowire Transistor: Complementary Metal-Oxide-Semiconductor Without Junctions', *Science of Advanced Materials*, 3(3), 477–482.

Colinge, J. P., Lee, C. W., Afzalian, A., Akhavan N. D., Yan, R., Ferain, I., O'Neill, B., Blake, A., White, M., Kelleher, A. M., McCarthy, B., and Murphy, R. (2010), 'Nanowire Transistors Without Junctions', *Nature Nanotechnology*, 5(3), 225–229.

Dambrine, G., Raynaud, C., Lederer, D., Dehan, M., Rozeaux, O., Vanmackelberg, M., Danneville, F., Lepilliet, S., and Raskin, J.-P. (Mar. 2003), 'What are the limiting parameters of deep-submicron MOSFETs for high frequency applications?' *IEEE Electron Device Letters*, 24(3), 189–191, Mar. 2003.

Gundapaneni, S., Ganguly, S., and Kottantharayil, A. (2011), 'Bulk Planar Junctionless Transistor (BPJLT): An Attractive Device Alternative for Scaling', *Electron Device Letter, IEEE*, 32(3), 261–263.

Kencke, D. L., Chen, W., Wang, H., Mudanai, S., Ouyang, Q., Tasch, A., and Banerjee, S. K. (1999), 'Source-side barrier effects with very high-K dielectrics in 50 nm Si MOSFETs', *Device Research Conference Digest*, 22–23.

Lee, C. W., Afzalian, A., Akhavan, N. D., Yan, R., Ferain, I., and Colinge, J. P. (2009), 'Junctionless multigate field-effect transistor', *Applied Physics Letters*, 94(5), 053511-1–053511-2.

Lilienfeld, J. E. (1925), 'Method and apparatus for controlling electric current', *U.S. Patent* 1745175.

Mondal, P., Ghosh, B., and Bal, P. (2013), 'Planar junctionless transistor with non-uniform channel doping', *Applied Physics Letters*, 102(13), 133505-1–133505-3, 133505.

Saini, G., and Choudhary, S. (2016a), 'Investigation of trigate JLT with dual-k sidewall spacers for enhanced analog/RF FOMs', *Journal of Computational Electronics*, 15(3), 865–873.

Saini, G., and Choudhary, S. (2016b), 'Analog/RF performance of source-side only dual-k sidewall spacer trigate junctionless transistor with parametric variations', *Superlattices and Microstructures*, 100, 757–766.

Sallese, J. M., Chevillon, N., Lallement, C., Iniguez, B., and Pregaldiny, F. (2011), 'Charge-based modeling of junctionless double-gate field-effect transistors', *IEEE Transactions on Electron Devices*, 58(8), 2628–2637.

Synopsys Sentaurus Design Suite (2020). [online] www.synopsys.com

Vinet, M., Poiroux, T., Widiez, I., Lolivier, I., Previtali, B., Vizioz, C., Guillaumot, B., Tiec, Y. Le, Besson, P., Biasse, B., Allain, F., Casse, M., Lafond, D., Hartmann, I.-M., Morand, Y., Chiaroni, I., and Deleonibus, S. (2005), 'Bonded planar double metal gate NMOS transistors down to 10 nm,' *IEEE Electron Device Letters*, 26(5), 317–319.

Index

For Product Safety Concerns and Information please contact our EU
representative GPSR@taylorandfrancis.com
Taylor & Francis Verlag GmbH, Kaufingerstraße 24, 80331 München, Germany